Hassina Hafida BOUKHALFA
Nacira CHOURGHAL

Mechanization and Agricultural Techniques

Hassina Hafida BOUKHALFA
Nacira CHOURGHAL

Mechanization and Agricultural Techniques

Preparing the soil to receive a crop

Imprint

Any brand names and product names mentioned in this book are subject to trademark, brand or patent protection and are trademarks or registered trademarks of their respective holders. The use of brand names, product names, common names, trade names, product descriptions etc. even without a particular marking in this work is in no way to be construed to mean that such names may be regarded as unrestricted in respect of trademark and brand protection legislation and could thus be used by anyone.

Cover image: www.ingimage.com

This book is a translation from the original published under ISBN 978-620-2-54825-0.

Publisher:
Sciencia Scripts
is a trademark of
International Book Market Service Ltd., member of OmniScriptum Publishing Group
17 Meldrum Street, Beau Bassin 71504, Mauritius
Printed at: see last page
ISBN: 978-620-3-15699-7

Foreword

This book sets out the main knowledge relating to the description, operation and use of agricultural machinery. We have limited ourselves to the agricultural operations that precede the installation of a crop and plan to deal with the following operations in the next works in the series "Mechanisation and Agricultural Techniques".

The work is organised according to a chronology linked to the order of the work to be carried out on agricultural land before crops are planted. Namely the stages of tillage and fartilisation, while presenting a detailed description of the agricultural machinery used in each cultivation operation and how it works. It also presented basic knowledge in terms of agricultural techniques and the different interactions with the environment and its constraints.

The book presents several alternatives as to the judicious choice of tool chains to be used according to the nature of the soil and the climatic conditions of the region. This is a very useful reference for farmers and agricultural advisers.

The content of this book is based on the training programme for agricultural engineers and is mainly dedicated to students in agricultural sciences to help them gain a wide range of knowledge during their university studies.

Table of Contents

GROUND WORKING

Introduction

Tillage includes all the cultivation methods that maintain a structural state of the soil favourable to the development of the plant. This to facilitate germination, growth and root installation. It also allows for the burial of crop residues (stubble), amendment (mineral and organic fertilizers) and the destruction of harmful vegetation.

Among the main objectives of tillage are the following:

1. Improvement of soil structure: this consists of reducing its tenacity and compactness, thus creating conditions more suitable for root development and facilitating the execution of other cultivation methods;

2. increase in permeability and porosity: this facilitates water infiltration, which has several effects: limiting stagnant water and surface runoff, a source of erosion, improving the balance between water and air in the soil thanks to the faster flow of excess water, and lastly promoting the replenishment of groundwater reserves ;

3. seedbed preparation: the crumbling of the clods creates an environment that places the seeds in the best germination conditions by facilitating their contact with soil particles and their moistening.

Tillage can also have many other effects, such as :

1. limitation of weed infestations,

2. limitation of water loss through evaporation,

3. levelling the surface of the land,

4. burying fertilisers, soil improvers or other substances such as pre-emergence herbicides.

Definition of the different soil conditions and its behaviour :

The consistency of a soil varies considerably with the moisture content of the soil. By gradually decreasing the water content of a soil sample it can be seen that the soil goes through several states in succession:

• **Liquid state with high water content.**

The soil behaves like a liquid. Its shear resistance is zero and it spreads when spilled. The soil grains are practically separated by water.

• **Plastic state :**

The floor is naturally stable, but as soon as stress is applied to it, it is subject to significant deformation, which is largely non-reversible without any significant variation in volume and without the appearance of cracks.

• **Solid state** :

The floor behaves like a solid, the application of a force causes only slight deformations. The transition to the solid state initially takes place with a reduction in volume or shrinkage, and then at constant volume, i.e. without shrinkage.

Estimation of soil consistency :

The consistency of the soil must be estimated on the plot just before working:

A few clods of earth should be taken from the work area and tried to crumble them between the fingers and compare the results obtained with the following table:

Behaviour	Tough	Friable	Semi-plastic	Plastic
Folding or sandy	Motte impossible to break	The clod crumbles easily	The lump is soapy	The clod becomes liquid
Intermediary	Motte impossible to break	The clump crumbles without sticking	The clump crumbles into a sticky mass.	The rootball is mouldable
Clayey	Motte impossible to break	The clump crumbles when sticking a little bit	The clod deforms and crumbles with difficulty.	The rootball is mouldable

Limits and Atterberg indices :

Liquidity limit

The liquidity limit (wl) characterises the transition from a plastic to a liquid state. It is the water content by weight, expressed as a percentage, above which the soil flows as a viscous liquid under the influence of its own weight. Formula for the water content by weight: Mass of water (g)/Mass of dry soil (g).

Plasticity limit

The plasticity limit (wp) characterises the transition from a solid to a plastic state. This limit indicates the maximum percentage of water by weight that can be used to work the soil and avoid compaction. Below this limit, the soil is crumbly or easily workable from an agronomic point of view. The plasticity limit is determined by shaping a small thread with the thin part of a soil on a flat, non-porous surface. It is defined as the moisture content, where the wire breaks at a diameter of 3 mm. A floor is considered to be non-plastic, if a thread cannot roll up to 3 mm, regardless of the moisture content of the thin part of the floor.

Liquidity index

$$I_l = \frac{W - W_p}{I_p}$$

Density index

$$I_d = \frac{e_{max} - e}{e_{max} - e_{min}}$$

Plasticity index

It measures the extent of the range of water content in which the soil is in a plastic state, **Ip = wl - wp**

According to the value of their plasticity index. Floors can be classified as follows:

Plasticity index	Degree of plasticity
0 < Ip < 5	Non-plastic (the test loses its meaning in this value range)
5 < Ip < 15	Medium plastic
15 < Ip < 40	Plastic
Ip > 40	Very plastic

Consistency index

This is a derived indicator:

$$I_c = \frac{w_l - w}{I_p}$$ where w=w sample normal

Choice of favourable conditions for the various soil works:

The distribution of soil types is based on their behaviour with regard to tillage implements. Soil behaviour depends both on the grain size (clay content) and the surrounding climate. For example :

• A soil poor in clay does not perform well in a dry climate, whereas it does in a wet climate.

• Also, clay rich soil does not stick to tools in a dry climate.

Therefore, the dominant characteristics of the plot must be retained and the following table must be estimated:

30% CLAY 15% CLAY	In the autumn, the soil stuck to the tools remains in mass without crumbling. This happens:	At the end of winter, is the floor swinging?	Floor behaviour :
	Never	Very often	Beating or sandy ground
	From time to time	From time to time	Intermediate behaviour
	Very often	never	Clay behaviour

Conventional tillage:

Conventional tillage comprises three distinct operations:

- Ploughing ;
- The resumption of ploughing or pseudo-ploughing ;
- Superficial ways.

Ploughing :

The purpose of ploughing is to loosen and sanitize the topsoil, which accelerates the speed of wiping and facilitates its exploitation by the crop roots. It also allows the crop residues to be buried and mixed with the soil so that they do not hinder the preparation of the seedbed, destroying harmful vegetation and certain parasites by burying them and facilitating the preparation of the seedbed. Seedbed quality and energy indices are based on :

- The physical and chemical nature of the soil to be worked.
- The depth of the topsoil.
- The time of the year (the climate).

Ploughing theory :

Théorie du labour

Ploughing theoretically consists in cutting a strip of land with a rectangular section ABCD, and turning it in a position A'B'C'D' placed at about 45°. This position may vary in practice depending on the ratio of width to ploughing depth.

The distance AB represents the ploughing **depth.**

The distance BC represents the ploughing **width.**

The plan determined by the AB line is called **wall** or sometimes **frayon.**

The plane determined by the BC line represents the **bottom of the line** or **gauge.**

The rectangular area ABCD represents the ploughing **section.**

Types of ploughing :

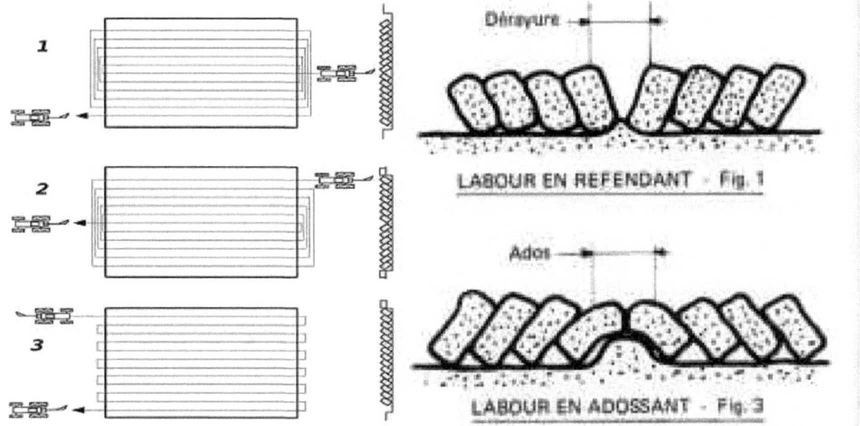

Ploughing methods in plan
1. Ploughing in plank mode with the boards leaning against each other.
2. Ploughing in planks while splitting.
3. Flat ploughing.

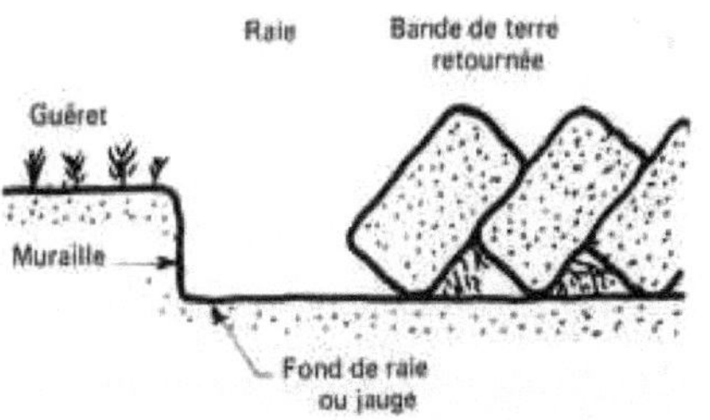

Ploughing can be carried out in two ways, depending on the ploughing pattern and the type of plough used:

- **flat ploughing, with** the strips of land always being thrown back on the same side. It requires the use of a reversible plough so that the direction of discharge can be reversed on a return journey;

- **ploughing in planks or ridges**. This is the only one that can be done with a simple plough that turns around the plot, and it can be done :

o **by slitting,** the strips being rejected towards the outside of the board (leaving a "scratch" in the centre of the board),

o **by leaning back**, the strips being rejected towards the axis of the board (leaving a "teenager" in the centre of the board).

A distinction can be made according to the depth of the work:

- **light ploughing**, from 10 to 15 cm, carried out in particular for the resumption of ploughing in spring,
- **medium ploughing,** from 15 to 30 cm, which is the most widespread, particularly for cereal crops,
- **deep ploughing,** from 30 to 40 cm, for deep-rooted crops (beet, alfalfa, etc.),
- Beyond 40 cm, **ploughing** is carried out in order to allow the cultivation of new land or to prepare the planting of orchards.

A distinction can be made according to the inclination of the strips :

- **upright ploughing,**
- **discarded ploughing,**
- **flat ploughing**.

PLOUGHING EQUIPMENT :

The equipment used for ploughing is the plough. Ploughs can be classified according to several criteria:

1. According to destination :
    - plough for ploughing in the fields.
    - Plough for ploughing in orchards.
    - Plough for vineyard ploughing.
    - Defensive plough.
2. According to the shape of the workpieces:
    - Plough with shares and mouldboards.
    - Disc plough.
3. According to the ploughing method :
    - Normal plough (turns furrow to one side only).
    - Reversible plough or pendulum plough (turns the furrow successively on both sides).
4. Based on ploughing depth :
    - Plough for shallow ploughing (08 to 15cm).

- Plough for normal ploughing (15 to 25cm).
- Plough for deep ploughing (25 to 35cm).
- Defensive plough (over 35cm).

5. According to the hitching mode :

- Dragged (pulled) plough.
- Semi-mounted plough.
- Mounted plough.

I. Ploughshare and mouldboard ploughs :

I. 1 Constitution :

The plough with share and mouldboard consists of the following components:

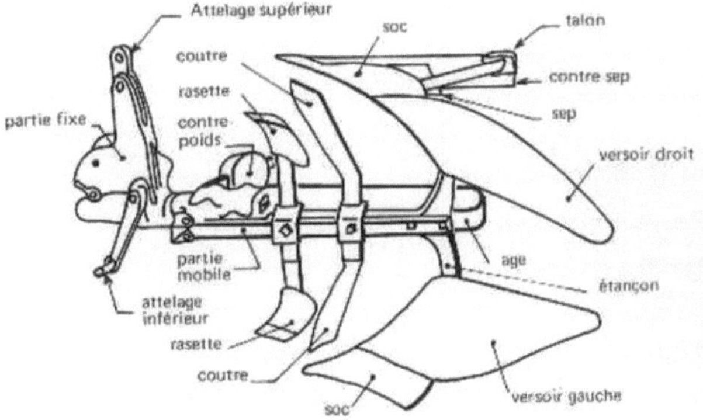

I. 1. 1. Working parts :

The "ploughshare, mouldboard" unit plays the main role in cutting the furrow by moving it through the soil. The furrow is compressed, moved and turned over onto the previously ploughed strip of land. The cutting in the vertical plane is carried out by the coulter.

I. 1. 1. 1. The seam :

It is a piece in the form of a knife or disc, its cutting part is arranged so as to cut vertically the strip of earth to be worked. It is therefore generally placed in front of the other working pieces. There are two types: **Straight beam** and **Circular beam**.

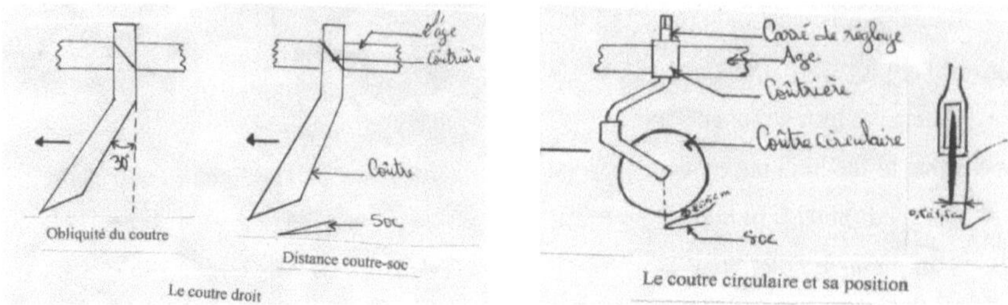

Le coutre droit — Le coutre circulaire et sa position

I. 1. 1. 2. The ploughshare :

It is a generally trapezoidal steel blade. Its role is to cut horizontally the strip of earth cut vertically by the coulter and to start lifting it in order to overturn it through the mouldboard. The width of the coulter is slightly less than the strip of soil so as to leave a part which serves as a pivot when turning over.

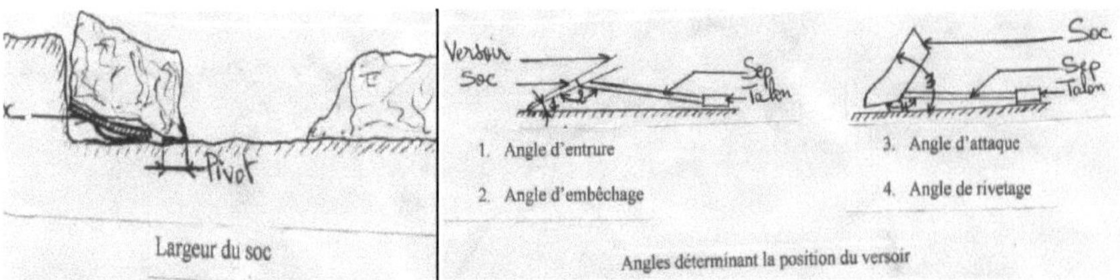

Its position is determined by four angles :

- **Angle of entry**: it forces the plough to constantly try to get into the ground. The value is 5° to 15°, but it can be as high as 25° for rapid ploughing.

- **Angle of attack**: it gives the cutting edge of the share an oblique angle to make it easier to cut the soil. It is 45° but its value can go down to 35° on certain bodies for rapid ploughing.

- Ploughing **angle**: it transfers the plough's support to two points, which increases stability and reduces wear. Its value is slightly less than 180°.

- **Riveting angle**: its role and value is identical to the angle of engagement but in a vertical plane.

I. 1. 1. 1. 3. The verso :

The role of the mouldboard is to turn over the strip of land previously cut by the coulter and the share. The mouldboards are made of triplex steel and their shape determines the way in which the land is turned over. This shape can be cylindrical, helical or mixed.

- **Helical mouldboard**: this shape allows the mouldboard to accompany the strip of land until the end of its turn, which makes it possible to obtain a well poured ploughing even in heavy soil. The soil is well loosened compared to the cylindrical mouldboard. It is mainly used for shallow ploughing.

• **Cylindrical pourer**: it initiates the turning of the strip of land which must then be completed thanks to the combined action of gravity and speed. It cannot be used for heavy soils or with very slow traction. On the other hand, the traction effort required is a little less and the loosening obtained is much better. It is also suitable for deep ploughing.

• **Mixed or universal verso**: it is cylindrical in its lower part, and helical over a greater or lesser length of its rear part. It combines the advantages of both forms and is the most commonly used.

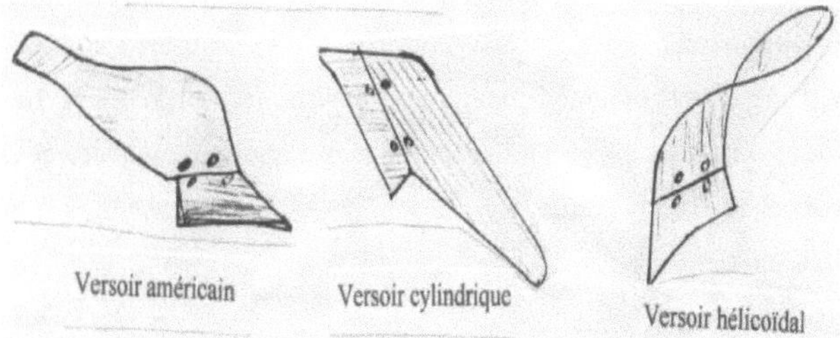

I. 1. 1. 4. The skimmer :

It consists of a real miniature plough body. It is generally placed in front of the other parts and fixed on the age. Its work consists in peeling the superficial part of the soil and sending it to the bottom of the furrow so that all the grass, manure or plant debris is perfectly buried, thus leaving a perfectly clean ploughing.

I. 1. 2. Supporting parts :

I. 1. 2. 1. Age :

It is the main supporting piece through which the traction of the plough is exerted. It supports the plough bodies by means of the plough legs. Its general shape is straight with round or rectangular cross-section.

I. 1. 2. 2. The stanchions :

Their role is to support the plough bodies. They are bolted to the frame. The connection between the plough beam and the frame includes a safety device based on a spring or shear rivet allowing the mouldboard-share assembly to overcome obstacles.

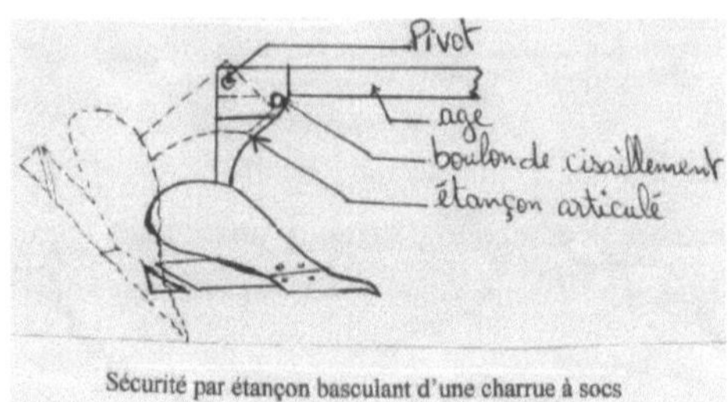

Sécurité par étançon basculant d'une charrue à socs

I. 1. 2. 3. On Sep :

The role of the sep is to connect the lower ends of the struts. It ends at its front part with the pallet which contains the holes for fixing the share and the bottom of the mouldboard.

I.1. 2. 4. the support wheel :

It is fixed to the age of the mounted plough and has the role of keeping the working depth constant by rolling on the stubble. Trailed ploughs are equipped with three wheels: stubble wheel, furrow wheel and rear wheel.

I. 1. 3. protective parts :

I. 1. 3. 1. The counter-sep :

It is a steel strip placed on the side of the stanchions and sep to protect them from the consequent wear and tear from rubbing against the wall.

I. 1. 3. 2. The heel :

Wearing part placed at the rear of the sep or counter-sep to prevent premature wear.

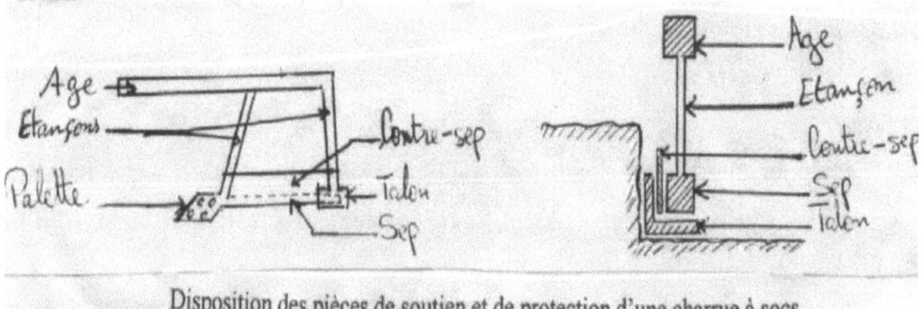

Disposition des pièces de soutien et de protection d'une charrue à socs

I. 2. Settings :

I. 2. 1 Adjusting the ploughing depth :

It is always obtained by varying the distance to the ground of the front part of the age, which has the effect of varying the angle of entry. The depth is increased by lowering the age and vice versa.

- On a trailed plough, a lever or ram varies the position of the gauge wheel (for the plough frame) and therefore the height of the plough frame.

- On a mounted plough, this adjustment is achieved by the height position of the lifting arms controlled by the tractor's hydraulic system.

I. 2. Adjusting the working width :

The only adjustment that the user usually makes is the width of the first furrow, which is equal to the distance between the inside of the tractor tyre and the point of the share.

I. 2. 3 Adjusting the tip inclination :

This adjustment should complement the width adjustment without replacing it. The tip inclination can be seen as a change in the orientation of the ages in relation to the direction. It causes the plough to work more to the right or more to the left as the case may be. The offset obtained by pointing should never exceed 8 to 10 cm.

I. 2. 4. Tilt or plumb adjustment :

The plough is plumb when the plough legs are perpendicular to the ground being worked. This position varies with the ploughing depth and should be adjustable. This adjustment is always achieved by swivelling the plough frame around its longitudinal axis. In the case of polyshare ploughs, the swivelling takes place around a central axis. Plumbing faults cause a variation in depth between the right and left plough bodies, which results in irregular ploughing, often referred to as "twin" ploughing.

I. 2. 5 Adjusting the edging :

This adjustment is usually of interest for trailed ploughs, it is obtained by lateral displacement of the coupling point. It is correct when the furrow wheel of the plough passes regularly through the angle formed by the wall and the furrow bottom. A maladjustment causes the wheel to move into the middle of the furrow or tries to mount it on the headland.

I. 2. 6 Adjusting the heel adjustment :

The correct plough bottoming of a plough is visible by the trace left by the heel in the furrow bottom, it should be marked without exaggeration. A plough which is too close to the furrow bottom lacks stability on the front axle and therefore tends to oscillate from right to left and vice versa.

I. 2. 7 Seam adjustment :

This setting is correct when the ploughing wall is perfectly neutral and vertical. When the coulter is too deep, the wall is uneven and toothed. When the coulter is too far out, the wall is stepped.

I. 2. 8 Adjusting the skimmer :

Insufficiently adjusted skimmers make it difficult to bury weeds and stubble. Setting the skimmer too deep results in poorly stepped furrows. In very hard soil, the skimmer is removed.

II. Disc plough :

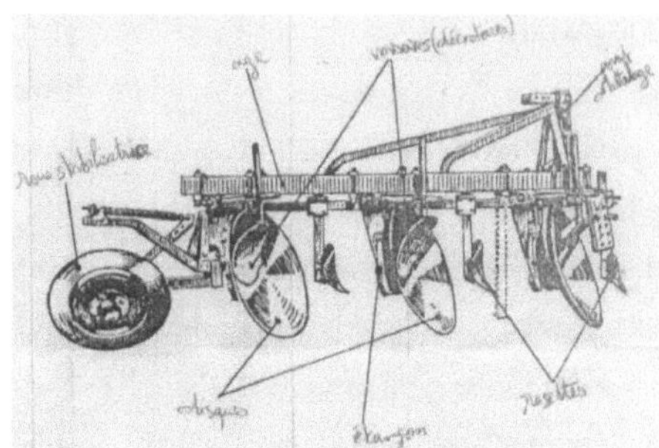

The disc plough can be trailed, mounted or semi-mounted. However, the most suitable solution is the trailed one because of the high weight of the discs. They should be used when cultivating on rocky (stony) or root-clogged soil. They are also used on land covered with vegetation and on heavy and compact land.

As far as the working principle is concerned, all the working parts of a share plough are replaced here by a single disc whose double-angled mounting enables it to cut the strip of land into an elliptical section and then turn it over.

I. 1 Constitution :

I. 1. 1. Working parts :

I. 1. 1. 1. The disc :

It has the general shape of a spherical cap. It must be very hard enough to resist wear and tear, and also flexible enough to prevent breakage. Its diameter usually varies from 610 to 815mm and its thickness from 6 to 75mm. The hub of the disc is fixed to its stud in such a way as to allow the angle of attack or entry to be controlled. The position of the disc is determined by two angles :

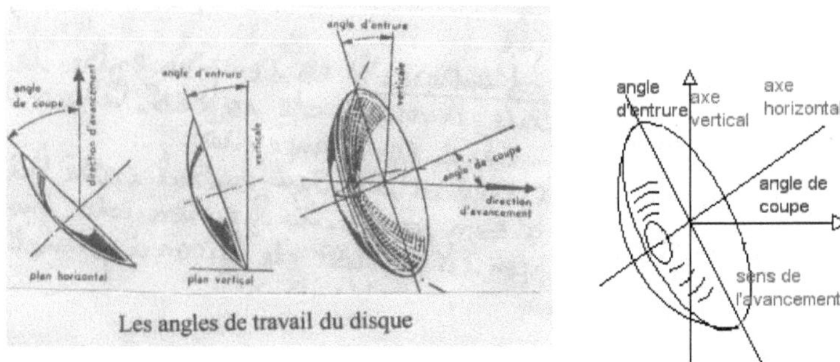

Les angles de travail du disque

- **Angle of attack**: also called cutting angle. It is between the feed direction and the plane of the disc. Its value varies from 30° to 55°.

  - **Entry angle**: it is between the plane of the disc and the vertical, varying from 15° to 25°. This variation favours penetration in soils with a hard consistency.

I. 1. 1. 2. The scraper :

Also known as a versoir or décrolette rasette. It is a curved piece that prevents the earth from sticking to the disc, thus making a complete turn with it, which improves turning and prevents jamming.

I. 1. 1. 3. The skimmer :

It is comparable to the one used on the plough and has the same function.

I. 1. 2. Supporting parts :

I. 1. 2. 1. Age :

Comparable to that of share ploughs. It can be common placed obliquely or multiple.

I. 1. 2. 2. The stanchion :

It ensures the junction between the disc and the age. Its position on the age can be fixed or adjustable in distance, and sometimes in orientation (action on the angle of attack and the angle of entry).

I. 1. 2. 3. stabilising wheel :

It is placed at the back of the plough. Its role is to brace the plough, which always tends to move away from the ploughing under the effect of the resulting forces due to the obliquity of the discs.

It rolls in the last furrow and its position is oblique in the opposite direction of the discs to better resist lateral slippage.

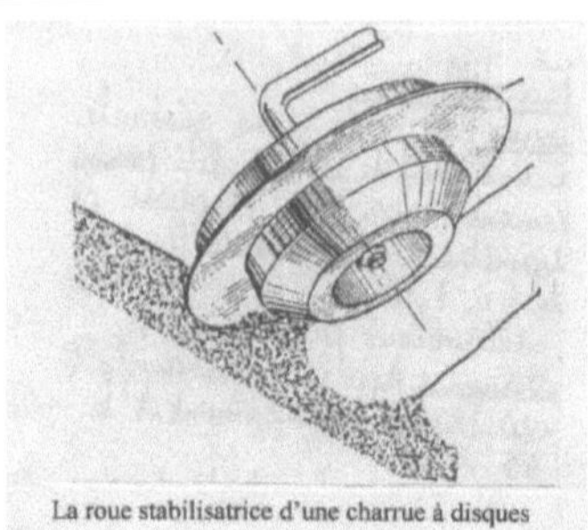

La roue stabilisatrice d'une charrue à disques

II. 2. Settings :

II. 2. 1. Depth adjustment :

This variation can be obtained :

- **By adjusting the entry angle**: reducing the **entry angle** results in an increase in depth. On the other hand, it cannot be increased too much, otherwise blockages will occur.

- **By lateral displacement of the frame**: this displacement must be horizontal, which means that the front and rear of the plough must be **moved** simultaneously.

On trailed ploughs depth adjustment is achieved by height adjustment of the wheels, and on mounted or semi-mounted ploughs by lifting the tractor.

II. 2. Width adjustment :

This adjustment can be obtained by varying the angle of attack or sometimes by moving the struts over the age.

II. 2. 3. Secondary settings :

These settings include: edging, heel and plumbing. They are strictly comparable with those of share ploughs, except for edging, where the lateral displacement of the coupling point is accompanied by a corresponding orientation of the furrow wheel.

PLOUGHING EQUIPMENT

I. Stubble cultivators :

They are designed for very shallow ploughing, favouring the emergence of weeds to help destroy them later. They can be either coulter or disc type.

Disc harrow :

I. 1 Constitution :

The discs of a stubble plough are similar to those of a disc plough. Their diameter varies from 560 to 610mm with a thickness of 4 to 6mm. Their main characteristic is that they have a zero entry angle, thus making it possible to mount them on the same shaft.

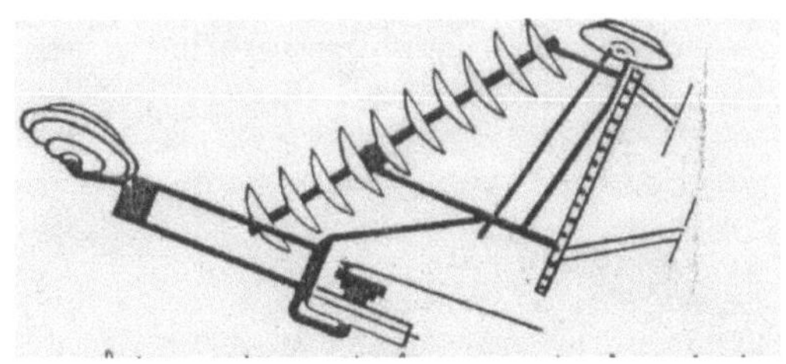

For mounted models, the frame is limited to a frame supporting the disc shaft and accommodating the three coupling points. They are also equipped with a stabilising wheel to balance the axial thrust of the discs. For trailed models, the frame is comparable to that of a disc plough. It is usually equipped with three heavy, sharp wheels, two of which are oblique to resist the radial thrust of the discs.

I. 2. Settings :

- Depth adjustment is obtained on models with a wheel limiting the entry, or directly by the position of the hitch arm. And on trailed models by bringing the discs closer to the ground or by changing the angle of attack by changing the orientation of the wheels. Where penetration is insufficient, ballast weights can be used.
- The edging is usually adjusted by deforming the coupling triangle.

II. Disc cultivators :

Disc cultivators ensure good soil crumbling and good weed control. They can be used for stubble ploughing or burying green manure, but penetration is often difficult in dry conditions.

II. 1 Constitution :

The working parts are discs similar to those of a stubble cultivator, with a smaller diameter (450 to610mm). The angle of entry is zero and the angle of attack varies according to the type of setting used. These machines use from 2 to 4 elements each consisting of a set of 4 to 15 discs mounted on the same shaft. The discs are mounted on these elements by inverting the concavity in order to balance the lateral thrusts, thus saving the work wheel assembly.

Disc cultivators are almost always equipped with angle frames to take up depth-dependent overloads.

II. 2. Different types : According to the arrangement of the elements we distinguish:

II. 2. 1. Single cultivator :

The single sprayer consists of two disc lines arranged in an open V-shape on the front side, the line of traction is centred on the tip of the V. The diameter of the discs varies from 450 to

560mm, and their number from 8 to 20. The disadvantage of this arrangement is that the small width of the V-tip is not worked, which often means that a cultivator tine has to be fitted at this point.

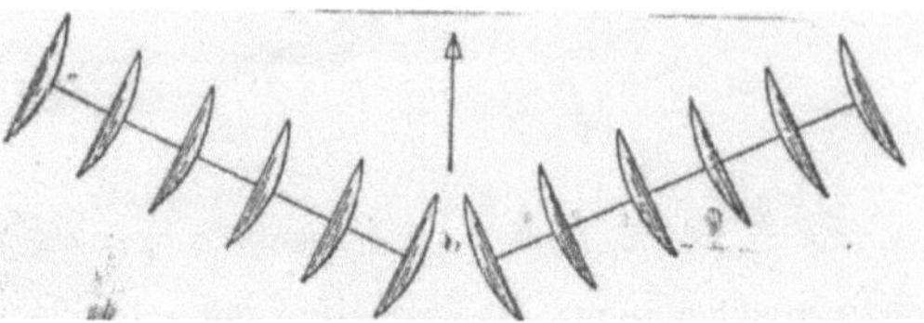

II. 2. Double or Tandem Cultivator :

They are made up of a combination of two simple inverted sprayers. The soil is therefore worked twice: the first set of discs inverts it on one side and the second one brings it back into place. The number of discs varies from 16 to 40.

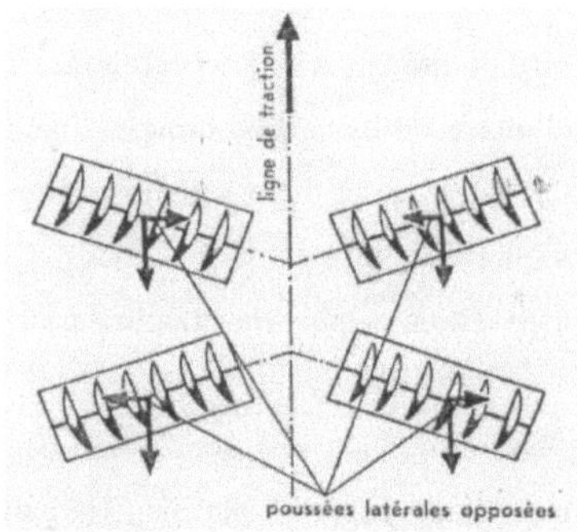

II. 2. 3. OFFSET tiller or Cover-Crop :

These machines have two rows of discs placed one behind the other, allowing the soil to be worked twice. In order to balance the lateral thrust of each row of discs, a higher angle of attack must be used for the rear element, which works already loosened soil giving less thrust. The diameter of the discs varies from 510 to 810mm and their number from 8 to 44. The high weight of these machines allows them to replace a stubble cultivator.

In addition, they leave no central part untouched like the single or tandem model.

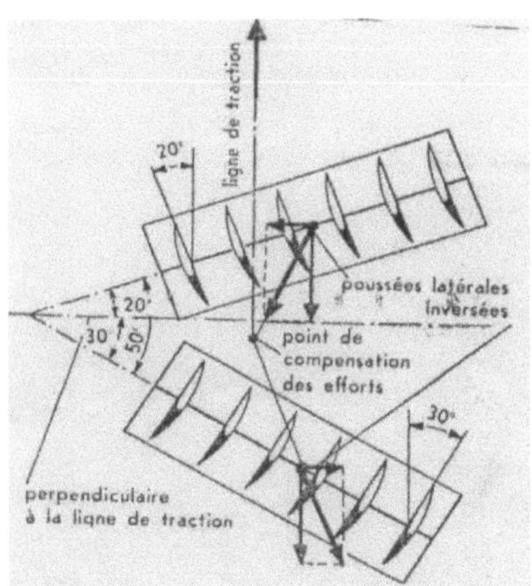

II. 3 Settings :

- The maximum depth is mainly related to the diameter of the discs used. For a given disc diameter, the depth varies with the weight of the equipment, which may be weighed down by the loads placed in the frames provided for this purpose. On appliances with support wheels, it is also possible to limit the infeed by adjusting their height.

- Flanging is achieved by moving the coupling point sideways.

III. Tine cultivators :

These machines allow the soil to expand deeply and aerate well, and also effectively destroy weeds and crumble clods. They also allow the superficial incorporation of fertilizers and grass. Tine cultivators alone have the property of not compacting the soil at depth.

III.1 The shares :

Different share shapes can be mounted on the same struts depending on the work to be carried out.

III. 1. 1. scarifying coulters: Scarifying consists in loosening the soil deeply without trying to work the surface much or to destroy weeds. These coulters are generally long and narrow with often two symmetrical ends that double the working life by reversibility.

III. 1. 2. Digging up ploughshares: Digging up is essentially aimed at destroying weeds accompanied by surface loosening. The coulter is therefore wide and has a low entry height since it works in an almost horizontal position at a depth just necessary to cut the weeds.

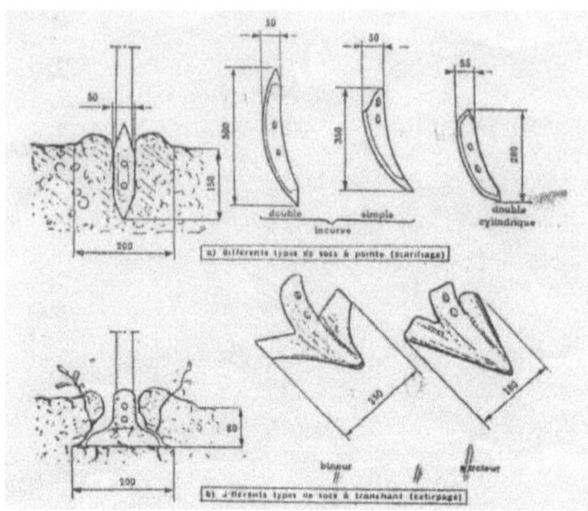

III. 2. Different types of struts :

III. 2. 1 Rigid stanchion combined with a spring: The stanchion is held in place by one or two spiral springs. This solution allows the stanchion to work safely by moving away from an unforeseen obstacle. Another realization consists in using powerful leaf springs maintaining strong rigid teeth for very deep work.

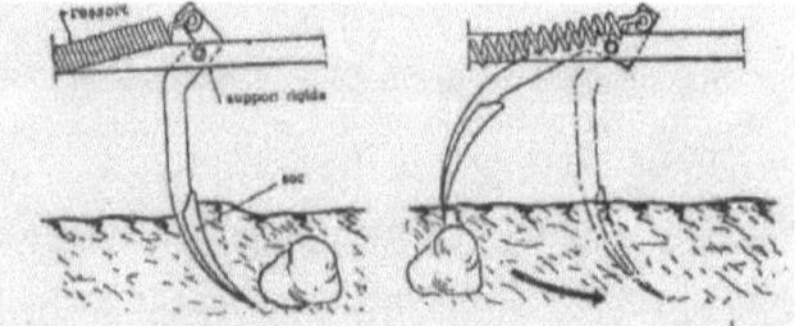

III. 2. Flexible flat steel stanchions: These stanchions are flexible over their entire length due to their thinness and the nature of the steel. Tools equipped with these types of props are often called Canadian Cultivators or Vibratory Cultivators. They carry out a medium-depth work with a high degree of loosening and infrequent blockages thanks to the great flexibility of the props, which are animated by strong vibrations in the soil.

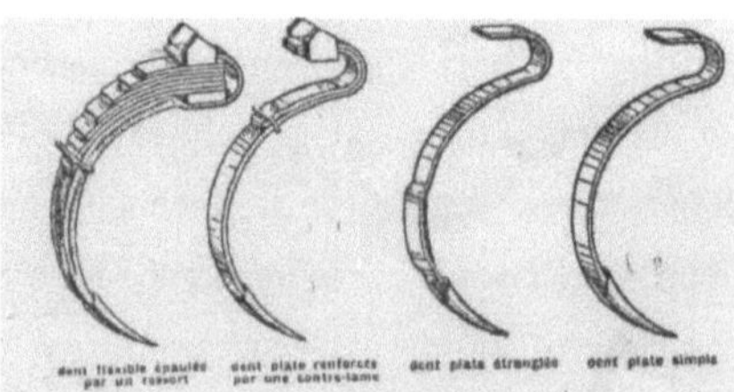

III. 2. 3. square steel buckle stanchion: This shape combines the great flexibility due to the double winding of the tooth. Cultivators equipped with it are suitable for dry and hard ground and for deep work.

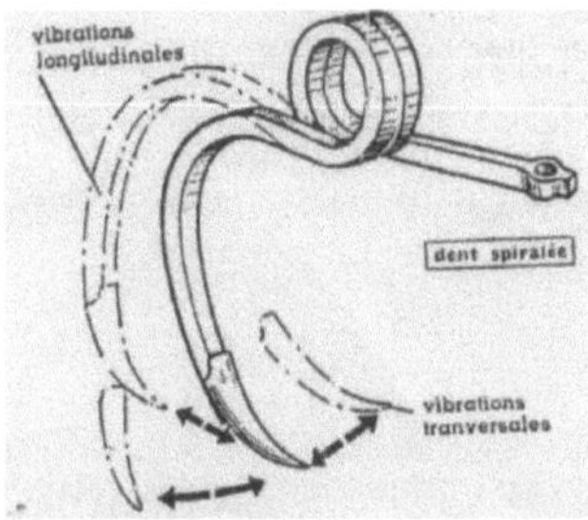

III. 2. 4. counter bent stanchion: This stanchion is very flexible and allows light work to be undertaken. Its shape is characterised by the counter-bend formed at ground level which forces the weeds to come to the surface where they dry out, while avoiding clogging and excessive rising of damp earth.

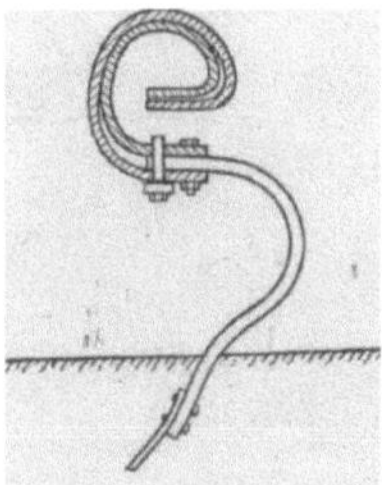

III. 3:

The frame is always designed in such a way as to allow the teeth to be arranged in such a way as to limit tamping as much as possible, by spacing out the neighbouring teeth over several crossbars. The appliances with a single crossbar are equipped alternately with long and short teeth to obtain the desired offset. The tine spacing is approximately 18 to 20cm, but can be as low as 10cm on some finishing vibra shapers.

III. 4 Settings :

- Attachments are depth-adjusted by the position of the lifting arms or by depth-limiting wheels that can be raised or lowered.

- Some machines are equipped with a rolling cage harrow. The depth of the cultivator elements is controlled by a chain connected to the harrow and fixed on a variable position bracket. This arrangement requires the use of the lift in floating position and allows excellent depth control while ensuring good seedbed compaction thanks to the weight transferred to the rolling cage harrows.

IV. Chisels :

The chisel can be considered as a giant tine cultivator. Its superficial use makes it impossible to differentiate its action from that of the tine cultivator. But its specific use implies a minimum working depth of 20 to 30 cm, as well as a working speed of at least 8 Km/h. This double condition leads to a general bursting of the soil.

IV.1.   Constitution :

IV. 1. 1. THE BUILDING :

Depending on the model, it has two or three large cross-section crossbars to which the struts, the lower hitch pins and the push bar support bracket are attached. In case of insufficient response of the hydraulic lift, it is possible to equip the device with a depth control wheel.

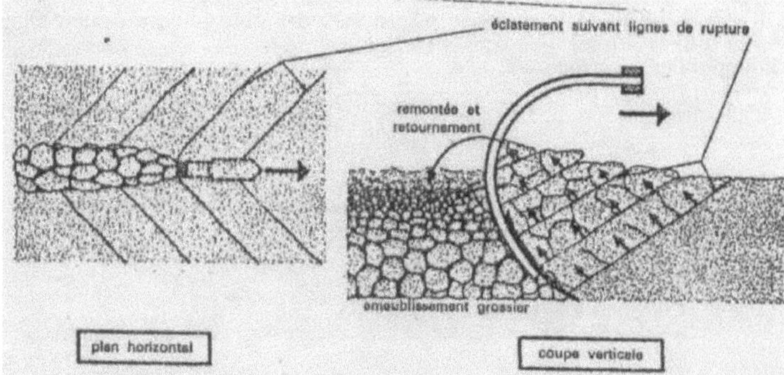

IV.1. 2. The stanchions :
They are judiciously distributed on the sleepers to avoid blockages. Their tracks on the ground are 30 to 40cm apart. Their height is such that the clearance under the frame varies from 65 to 100cm. Their most common general shape is close to a semicircle.

IV. 1. 3. The coulters :
Several shapes can be mounted depending on the case.

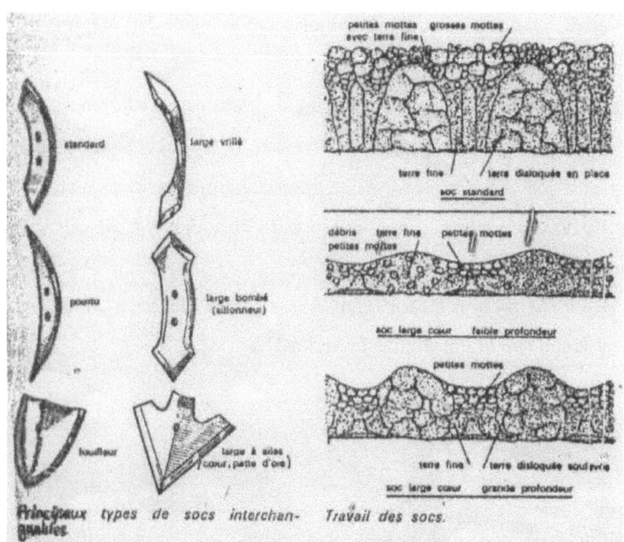

• **Reversible upright shares:** can be used for deep work in dry soil.

• **Twisted coulters:** they are used superficially to take advantage of the local turning of the soil resulting in a slight ridging of the surface and good incorporation of harvest residues.

• **Goose foot ploughshares:** are reserved for weed destruction work on already loosened land.

1.1. V.   Rotary cultivators :

These devices are also known as "Strawberries". The working part consists of a horizontal shaft with a variable rotation speed from 130 to 230 rpm, borrowed from the tractor's power take-off shaft, on which are threaded a number of flanges, each bearing generally 6 knives (or spades) of variable shapes, and separated from each other by spacers keeping them at a distance of about 25cm.

These devices can be carried or semi-carried.

V. 1 Different shapes of spades :

• **Straight spade:** It is used on clean and relatively light ground or to destroy dry clods.

• **Angled spade:** It provides good loosening in all types of soil by allowing plant debris to be easily buried.

• **Helical spade:** It adapts well enough for different types of work.

• **Double angled spade:** This model is used for bush clearing or land clearing work.

• **U-shape spade:** This shape is very solid, allowing work in hard and dry ground.

V. 2 Settings :

- **Depth:** It varies from 0 to 25cm, the adjustment is made by varying the height of the wheels or skates.

- **Furnishing :** The fineness of the soil spraying depends essentially on :

- The tractor's forward speed, to which the loosening is inversely proportional.

- The speed of rotation of the tools to which the loosening is proportional.

In particularly difficult conditions, it is possible to consider two passages and cross if the shape of the part allows it.

V. 3 Safety: When the motion transmission has a safety clutch, it should be adjusted very gradually so that it can slip at the slightest overload.

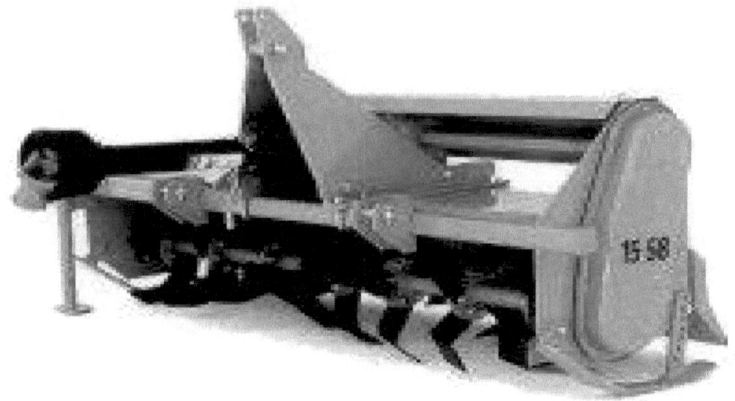

SHALLOW TILLAGE EQUIPMENT

I.Harrows :

Harrows work at shallow depths to achieve surface loosening, levelling of the ground, and destruction of weeds when they are young enough. They work in widths of 2 to 5m for mounted solutions and reach widths of 15m for trailed solutions. They can be differentiated by the shape of their working parts as well as by the possible movement that can be imparted to them.

1. Trailing harrows with fixed teeth :

- **Z-shaped harrow:** This is the most common form, it is generally made up of a variable number of elements called "compartments"; each of them generally includes 5 crosspieces and 3 to 5 arrows. With the teeth being fixed at each intersection, a compartment can have 15 to 25 teeth. Each tine is generally arranged one edge forward so that it is never on the same crossbar as the one working on the adjacent line to avoid jamming. Each compartment is connected to its neighbour at the front by the drawbar and at the rear by a balance or coupling bar.

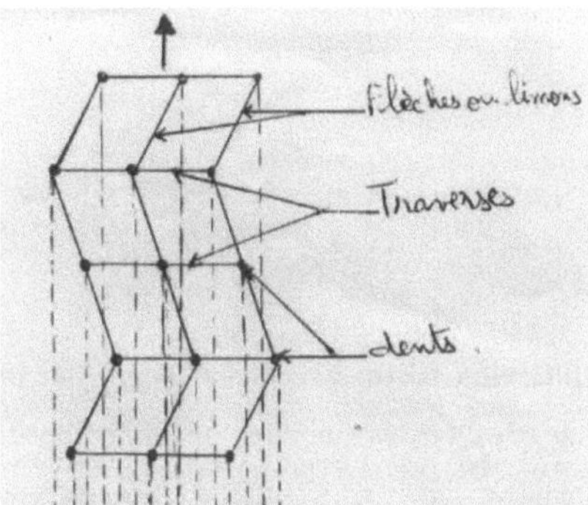

- **Flexible harrow:** This model does not have a differentiated frame because the tines are connected to each other to form a real flexible mat. This arrangement makes it possible to make very light harrows that are interesting for very superficial work, weeding a standing crop, working on ridges without flattening them...etc.

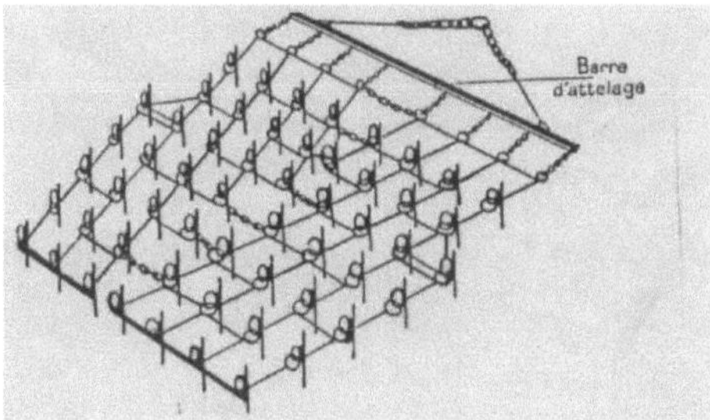

- **Rotary harrow:** This is a circular model suspended from a bracket adapted to the three-point linkage of a tractor. It is placed slightly obliquely in relation to the ground and automatically starts to rotate as it moves forward, assisted by an adjustable counterweight. This type of harrow can be interesting in shrubby crops. Moreover, the rotation avoids blockages.

- **Vibrating harrow with flexible teeth:** The fine and very flexible teeth, 30 to 40cm long, are distributed in staggered rows on two or three angles perpendicular to the direction of travel. These instruments achieve a very good surface loosening thanks to the very important vibrations of the teeth. In addition, they can also be used for light hoeing.

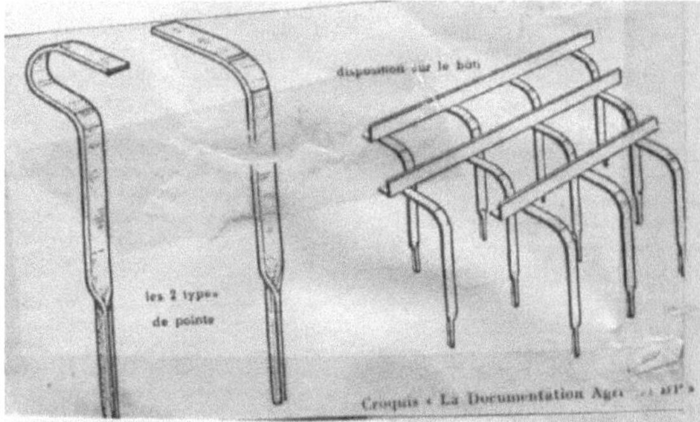

2. Harrows with rolling elements :

This category includes appliances with elements that rotate due to traction.

• **Rolling star harrow** : Consisting of one or two square shafts arranged transversely, on which are threaded cast iron elements with 5 points arranged in a helix to regulate the work. This device performs a good surface loosening even on hard ground, but causes a greater soil compaction than that of a harrow.

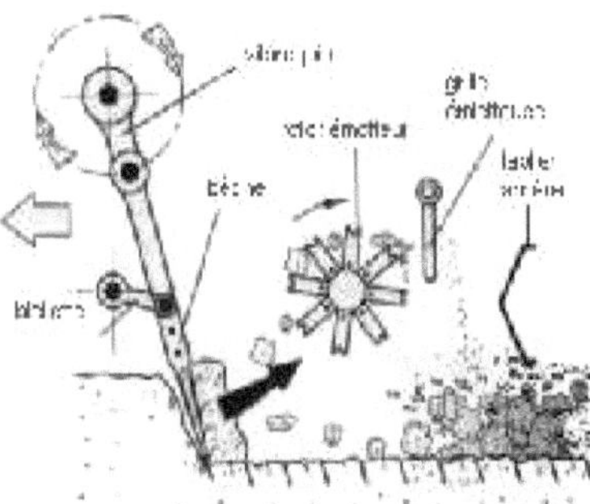

- **Spade harrows**: the working parts are made up of stars with four concave branches mounted on obliquely arranged trees. the number of successive trees (2 to 4) determines the intensity of the final work. This machine has good penetration qualities, it can perform very good surface loosening at a fairly high speed.

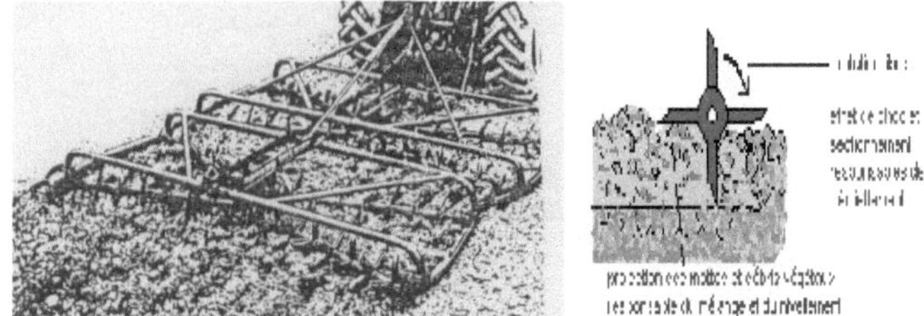

- **Rolling cage harrow**: the working elements are cylindrical cages with asperities on the outside. The movement of these cages makes it possible to fine-tune the surface loosening and to level the ground surface. These qualities are only truly enhanced in pre-soil, which is why these machines are usually combined with flexible tine cultivators, where they complement the action of seedbed preparation.

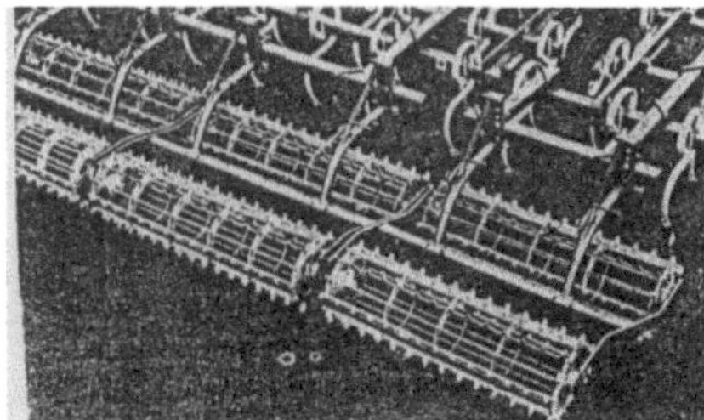

3. Harrows ordered :

These harrows are driven by the tractor's power take-off shaft. Depending on the movement imparted to the tines, two main families of tools can be distinguished:

• **Alternating harrows**: they are equipped with vertical tines of 20 to 30cm fixed on two or four bars arranged perpendicular to the forwarding blades and driven by a lateral translation movement. The combination of these movements and the feed results in a thorough loosening of the surface. With these machines there is a risk of forming a wet work surface.

• **Rotary harrows**: elements rotating around vertical axes usually have two tines and the neighbouring rotors rotate in opposite directions. A cage roller at the rear completes the work by levelling and reconsolidating the soil and gives better depth control.

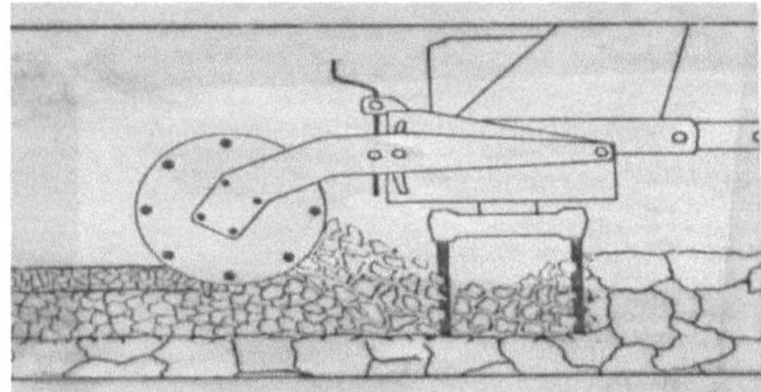

II. Hoeing machines :

The hoe is a tool designed to carry out very superficial soil maintenance and hoeing work in crops arranged in rows (beet, vines, etc.). The primary objective is to destroy weeds. However, by fragmenting the shallowest part of the soil, which will then dry out considerably, a discontinuity is created in the flow of water from the soil to the surface: hoeing thus also contributes to preserving soil water.

III. The rolls :

These tools are used to compact the soil or loosen it superficially by destroying clods. All the rollers allow this double action to be achieved, but each model has a dominant feature oriented towards one or the other.

1. **Smooth roller:** it is made up of a number of elements (or balls) allowing it to rotate without digging into the ground. Each element has a diameter of 40 to 60cm and a length of 50cm in general. The disadvantage of these devices is that they often cause a continuous crust to form on the surface of the soil, which hinders lifting in dry periods.

2. **Corrugated roller:** its structure and weight are similar to the smooth roller. The difference lies in the use of highly ribbed sheet metal. This design weakens the crust that is formed and makes it easier for plants to pass through.

3. **Skeleton roller:** it is made up of a series of narrow elements that are not joined to each other to avoid the continuous crusting of the soil and better destroy clods. This model tends to pack more than the others and its weight is a little higher.

4. **Croskill roller:** made up of a succession of cast iron elements mounted next to each other on the same shaft. Each element is notched on its periphery and on its side faces to better destroy clods. There are always two different diameters of about 5cm alternating on the same shaft; the larger ones have a central bearing much larger than the diameter of the shaft while the smaller ones have a normal bearing. This arrangement allows the small discs to roll normally while the large ones move forward in a jerky manner, creating friction that prevents clogging. The croskill is the most efficient but heaviest clod breaker and is only used for soil preparation.

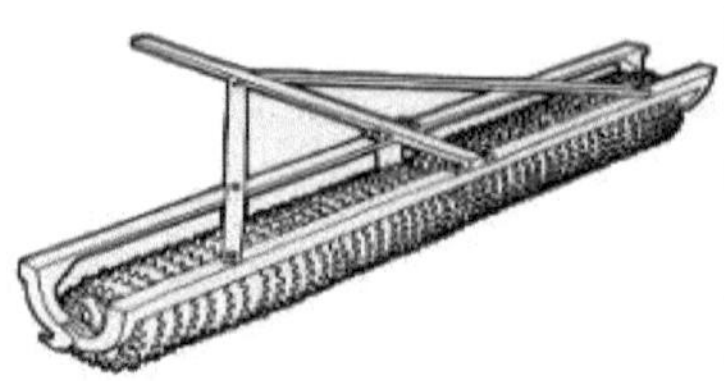 

5. **Star roller:** It consists of two or three shafts on which cast iron stars (diameter 25 to 40 cm) are mounted, rotating freely on their axis and interlocking in a helix so as to tangle up to sink the jams and perform a real self-cleaning of the tool.

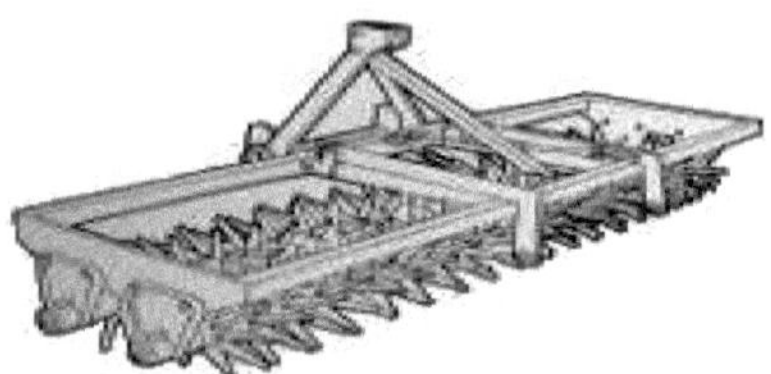

6. **Cultipacker roller:** This is also called a wedge ring roller and usually consists of two shafts working one behind the other. The elements have a sharp edge connected to the base cylinder by a curved part to stretch the clods. The front discs have a larger diameter than the rear discs. These rollers provide very good loosening.

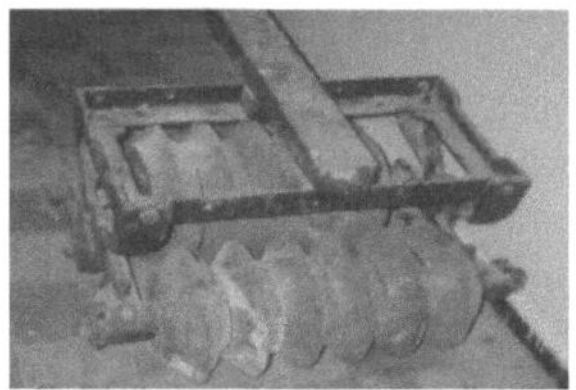

Minimum tillage :

This system is limited to shallow tillage to prepare the seedbed without loosening its depth.

No tillage :

This concept assumes that all tillage operations, including seedbed preparation and mechanical weeding measures, are eliminated. Only by strictly adhering to this principle will the positive action of soil fauna be fully exploited. This system implies a high dependence on chemical herbicides for weed control and the absence of any soil tilting operations.

WATER-SAVING CULTIVATION TECHNIQUES

I. Mulching or mulch:

A protective layer of crop residue will help to reduce high soil surface temperatures. It will improve the physical condition of the soil surface by promoting the activity of soil fauna such as earthworms and termites. This may help improve water infiltration. To establish and maintain wildlife activity, at least 2 tonnes of tailings are needed per hectare. The protection of the soil surface against the impact of raindrops is ensured by mulching the soil with 90-100% mulch, which represents between 6 and 8 tons/hectare of crop residues. These quantities cannot easily be produced by a single crop (except perhaps maize), so cover crops are needed. To manage these cover crops properly, appropriate tillage techniques are needed because conventional methods such as mouldboard or disc ploughs bury too much protective material. No-till' or direct seeding is the only way to leave residues. The large amounts of straw required cannot be produced or applied in semi-arid or arid climates, limiting the use of this system to sub-humid climates.

II. Dry farming or arido-culture:

Dry-farming is a cultivation method suitable for semi-desert regions, which makes it possible to grow plants, especially cereals, without the need for irrigation. The technique consists of ploughing very deeply to reach the wet layers of the soil and protecting the available water by breaking up the superficial clods of earth very finely. The land is only sown every other year, which helps to build up water reserves. The set-aside soil is thus ploughed several times to loosen the soil and to increase its capacity to absorb rainwater. This technique is practised in the Maghreb, in the drier plains of the interior, receiving less than 500 millimetres of rainfall. But dry-farming promotes soil erosion. Fallow land is attacked by wind and runoff, which tear away the topsoil. It is now in decline due to advances in irrigation.

III. Summer fallow:

This technique consists of leaving the land fallow for part of the agricultural season. It is then necessary to hoe the land during the normal growing season to avoid water loss due to weed growth. The purpose of summer fallow is to store water in the soil and to accumulate nitrate nitrogen in the soil for the next season. It is subject to three major constraints:

• The average annual precipitation must be sufficient to moisten the soil throughout the root horizon;

• the soil must be deep enough and have sufficient water retention capacity to retain the water available during the fallow period ;

• the cultivated species must have sufficient root development to use the stored water.

The type of soil influences the efficiency of fallow land, as it must have a water retention capacity that allows it to benefit from the additional water provided during the fallow period. In general, a soil that has a light texture of sandy loam or silty sand along the length of its profile will not be an adequate reservoir for storing moisture as would heavier textured soils.

IV. **Level pans**:

Surveys carried out on large areas with natural drainage on gently sloping ground have been effective in some semi-arid areas. Their purpose is to trap runoff water and store soil moisture for crops. These levelled ponds can vary from less than half a hectare to several hectares. They can be laid out in such a way that excess water falls from one level to the next. Grassy outlets are required for drainage and diversions are needed to reduce excess runoff water where it can harm crops. The amount of water collected in these ponds depends on the amount and nature of rainfall and the number of ponds in the system.

THE CULTIVATION SYSTEM :

Definition :

A cropping system is a theoretical representation of a way of cultivating a certain type of field. (CNEARC, 1989)

All the technical procedures implemented on plots of land treated in the same way (Sébillotte Michel). This concept therefore applies to the scale of plots that are used in the same way.

Elements of the cultivation system :

A cropping system is characterised by homogeneity in the way a crop is grown on a set of plots: same species, crop associations, same crop succession, same technical itineraries. It is a farming method common to a group of farms. Several cropping systems can be found on a single farm. A cropping system is :

- One or more species and varieties planted or sown ;

- A way of combining species in space: pure culture or association ;

- A given crop succession (a way of combining crops over time, sometimes cyclical: in this case we speak of rotation);

- One or more given technical itineraries "a set of cultivation practices ordered in time, applied to a crop or a combination of crops, from the preparation of the land to harvesting".

Rotations and rotations :

Crop **rotation:**   Distribution of crops in space.

Rotation: Succession in time of crop cycles on the same plot (distribution of crops over time).

On a plot in time :

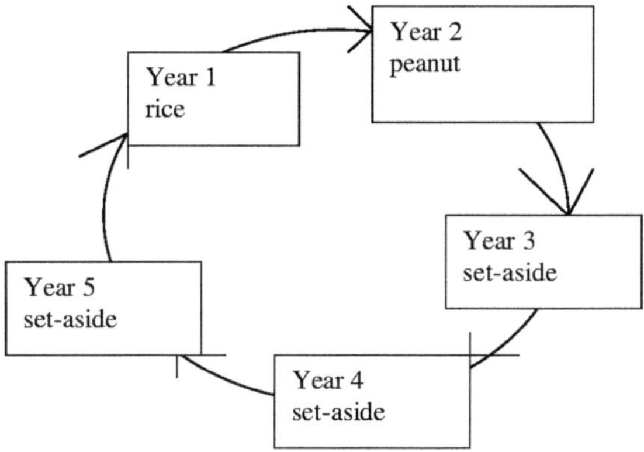

Over one year, in space :

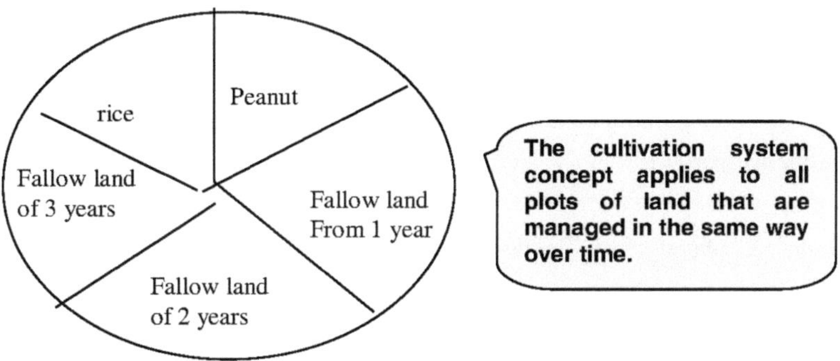

Several plots of a farmer "fit" into the same cropping system. This does not necessarily mean that in a given year all plots are at the same stage.

SPACE

	Plot 1	Plot 2	Plot 3
Year 1	Rice	Peanut	Fallow land
Year 2	Peanut	Fallow land	Rice
Year 3	Fallow land	Rice	Peanut

(TEMPS)

The same cropping system is practised in all three plots.

There is a relationship between the relative importance of a species in a rotation and its relative importance in the crop rotation. Example: Cotton // Sorghum // Millet // Millet: cotton represents 25% of the crop rotation, sorghum also and millet represents 50% of the crop rotation.

Example of rotation: Cereal / Onion / Tomato // Cereal / Onion / Tomato

Technical itineraries :

Definition :

The technical cultural itinerary is the set of cultural practices, ordered in time, applied to a crop or a combination of crops, from the preparation of the land to the harvest.

• For 1 cropping system: as many ITKs as there are crop cycles to follow, within a year and/or during a rotation cycle.

• In the case of associated crops, a single technical itinerary for the conduct of associated species

The organisation of the cultivation system :

Operating constraints :

To describe each of the operations :

Identify the tools, equipment and inputs used for each of the tasks.

How much seed is used? Where does it come from? Are fertilisers, herbicides, pesticides used? Is specific equipment (e.g. a sprayer) rented to apply them? What doses of products are used?

The establishment of the cultural calendar :

Intra-annual successions :

Definition: succession of crops on a plot of land over the course of a year.

Example: two cycles per year (cycle 1: M*N and cycle 2: Manioc)

		jan	fev	mar	Apr	May	jun	jul	August	sep	Oct	nov	Dec
Year 1	Corn Average cycle (90 days)				←			→					
	Niébé					←	→						
	Manioc	→							←				

A calendar of crop cycles: positions the cycles of the different cultivated species, from sowing to maturity, allowing a good visualization of the associations and intra-annual successions. It must be related to the ombro-thermal calendar of the region.

To sum up, therefore, a distinction is made when drawing up the calendar of cultivation operations :

· Periods in which operations are carried out

· The windows of time

· Work times

This flexibility or rigidity of the growing calendar is generally dictated by the seasons, as certain technical operations can only be carried out if optimal climatic conditions are met (beginning of the rainy season, for example). This information on the "flexibility" of the cropping calendar is essential for identifying peaks of work.

CULTURAL PROFILE :

Definition :

The crop profile is an agronomic diagnostic tool which, from the examination of a soil layer (1.50 m deep and 3 to 4 m long), enables principles of action to be drawn for agricultural practice (Hénin et al 1969). The strata thus defined are described in a very methodical way:

• Soil structure: internal state of the clods and how they are assembled with the fine soil.

• Soil water status.

• Distribution of the root system.

• Location and evolution of organic matter.

Study methods :

Once a representative area of the plot has been chosen, the dimensions and orientation of the pit must first be defined according to the question asked and the observation face must be chosen to protect it from any compaction or trampling during excavation. The pit must be perpendicular to the direction in which the soil is worked (which implies knowing precisely the technical itinerary of the plot) and sufficiently large and deep. For the observation of the profile, a few tools are essential: a digging fork, a knife with a blade of about 15 cm and rounded at the end, one or two double metres, and if possible a good bellows. Initially, we will try to locate the vertical and lateral partitions, which are essential for an understanding of the origin of the different structural states and a correct interpretation of the profile. As far as the structure is concerned, each stratum will then be described, characterising successively:

• The method of assembling the clods and the fine earth. State " o ", as in " open " for clods that are not very well bonded to each other and to the fine earth. State "b", as blocks, for decimetric clods with cavities. State "c", as "continuous" if there are no discernible clods.

• the internal state of the clods (each clod is broken in half to observe the fragmentation face).

   o Delta clods, flat face without asperity, no visible porosity, generally no roots or earthworms, resulting from severe settling.

   o Gamma clods, high roughness and porosity, no settling.

o Delta zero clods, intermediate state, flat but rough fragmentation face, low but not zero porosity, some roots and earthworm galleries.

o Phi clods, only if swelling clays are present, beginning of cracks, angular facets.

o Delta plus clods, perfectly flat face with traces of "folding", result from very severe settling.

It is also very interesting to observe the worm galleries visible on the observation face (empty and shiny galleries, or partially filled with excreta, presence of spherical cavities, etc.), and to count the open holes of worm galleries, on a horizontal plane such as the ploughing bottom. By relating all these observations (without forgetting the roots), it will be possible to progressively forge a diagnosis at the scale of the profile and therefore for the area studied on the plot.

FERTILISATION

Definition:

Fertilisation is the action of applying organic or mineral fertilisers, necessary for the proper development of plants. It can therefore be carried out in the form of humiferous (organic) or mineral (chemical) amendments. Most often, we speak of amendments when a fertilization effort is intended for the soil and we speak of fertilizer when the fertilization effort is intended for plants.

Objectives

1. Provide the plant with the nutrients it needs in kind, in the right quantity and at the right time.

2. Maintain soil fertility, especially the level of organic matter. By reasoning on the scale of the rotation. By taking into account technical, economic and environmental constraints.

I. LIMESTONE AMENDMENTS

Soil acidity and its disadvantages :

The consequences vary according to the level of soil acidification:

• degradation of the soil structure

• in non-calcareous soil, decrease in effective CEC (cation exchange capacity evaluated by a method with little soil modification), cation leaching. Soil fertility is reduced, elements such as phosphorus, potassium and magnesium are less available to the plant from a certain level of acidification.

• decrease in the biological activity of the soil.

• increase in the solubilisation of certain minerals which can cause toxicity (Al, Cu and Mn) if the pH drops too low (<5.5). These toxicities can affect the vine (in the case of young plants) but above all affect the life of the soil.

Reaction of soil and vegetation to acidity :

• field observations: natural flora, degradation of surface condition, more difficult degradation of organic residues.

• the water pH which is the pH of a suspension of earth in pure water in a given earth/water ratio (1/5 by volume French standard). This pH varies according to the season and there is no ideal pH.

• the S/CEC saturation rate: soil acidification results in a loss of exchangeable cations (Ca^{2+}, Mg^{2+}, K^+, Na^+) from the exchange complex and their progressive replacement by H^+ and Al^{3+} ions.

• buffer capacity: this notion reflects the soil's greater or lesser ability to moderate pH variations.

Correction of soil acidity :

• pH water < 6.2 => correction required

• 6.2 < pH water < 7.0 => maintenance

• pH > 7.0 => no addition

Dose calculations are made taking into account the difference between the desirable state and the current state and the buffering capacity.

Role of calcium in relation to the plant :

Mainly active in the form of carbonate, Ca2+ is a plant food that contains from 0.14 to 45 ‰ CaO (CaO = 1.4 Ca), depending on the species, the nature of the organs, the age; seeds, fruits, roots and tubers are less rich in Ca than leaves. The calcium content of leaves increases with age.

It is exceptional that the soil does not provide a suitable calcium supply for the plant, as even very poor soils contain sufficient quantities to ensure the plant's food requirements, between 25 and 100 kg of calcium per hectare. However, calcium is mainly responsible for the pH and the efficiency of the clay-humic complex in the soil: this is its essential role as a soil improver. It has a positive effect on :

-Structural stability, allowing a stable granulation of the particles, facilitating the passage of air and water and the penetration of the roots.

-PH, which is generally optimum around 6.5 (and even 7.5 in clay-silt soils on the plateaux). pH values above 8 are generally unfavourable to vines and certain fruit trees such as pears, or to certain annual plants (soya, sorghum, lupine) due to an excess of active limestone leading to iron chlorosis.

-acidic earth micro-organism activity.

-the availability of the soil in certain mineral elements, as it promotes the mobility of K+ and maintains PO43- ions in assimilable forms.

The full effectiveness of mineral fertilisers can only be achieved on soil in good calcium condition.

Main causes of calcium loss in the soil :

They are due :

• **The acidifying nature of the fertilisers**: 100 kg of NPK (17-17-17) results in a loss of 21 kg CaO, 100 kg of ammonium nitrate of 33 kg CaO, and 100 kg of urea of 46 kg CaO.

• **To the harvest :**

Culture	CaO losses in Kg/ha
Cereals	40-50
Beetroot, potato	90-120
Alfalfa	250-300
But silage	50
Fodder grasses	60

• To **rainwater entrainment**: Ca^{2+} is easily entrained by rainwater loaded with carbon dioxide and sulphur. In total, these losses vary in intensive cultivation from 400 to 800 kg CaO, or equivalent units of CaO (neutralising value).

Evaluation of the lime requirements of soils :

It is determined in the laboratory by measuring the pH and the saturation rate of the C.E.C. (cation exchange capacity). It is calculated by taking into account the quantities of calcium necessary either to maintain the physical properties of the soil (higher than those necessary to obtain a water pH of 6), or to maintain a satisfactory biological activity of the soil (equal to those necessary to obtain a water pH of 6).

A soil in a state of good productivity should contain about 3 ‰ of exchangeable CaO if it is sandy, 2 to 3 times more if it is clayey or humus (depending on the E.C.C.). If the analysis or certain visible signs (poor structure, slow decomposition of organic inputs, difficult implantation of alfalfa) reveal a lack of lime, the pH is gradually raised.

To avoid a blockage of trace elements, most of which (except molybdenum) are less assimilable by plants in a highly alkaline environment, it is preferable to make moderate additions of limestone soil improvers, but repeated every 3 to 5 years.

Maintenance liming, each time raising the pH by a quarter or half unit, should become the rule.

To raise the pH by half a unit, the following quantities/ha of soil improvers are required on average:

	Quicklime	Limestone (50% CaO)
Sandy soil	400 to 1000 kg	800 to 2000 kg
Limons	800 to 2000 kg	1700 to 4400 kg
Clay or humus soils	1300 to 3400 kg	2600 to 6800 kg

The different limestone amendments and their instructions for use :

There are raw products (limestone, chalk, dolomite, marl) and faster acting cooked products (lime). A raw product in granulated form must be incorporated into the soil after spreading, at the risk of losing all its effectiveness after recarbonation. Basic soil improvers are characterised by :

• **Their neutralising value (VN):** This is the quantity of calcium oxide CaO with the same neutralising capacity as 100 kg of the product under consideration. For example, the neutralising value is 100 for pure lime CaO, and 56 for pure calcium carbonate. Magnesia has, for the same weight, a greater neutralising capacity than calcium oxide: 1 MgO is equivalent to 1.4 CaO. For example, a dolomite containing 30% CaO and 21% MgO will have a neutralising value of 30+21 x 1.4 = 59.

• **Their carbonic solubility**, characterising its rapid action in the soil. We thus have 3 categories of products: fast action (carbonic solubility $\geq$ 50), medium fast action (from 20 to 50), slow action (< 20).

- **Their finesse.** More fineness allows a better speed of action. The amendments are called :

-sprayed if 80% of the product passes the 0.315 mm sieve and 99% minimum at 1 mm,

-grinded if 80% of the product passes 4 mm,

-crushed or raw if their granulometry is larger than that of the crushed products.

Nature of the limestone and magnesian amendments :

Composed mainly of calcium or magnesium oxide or carbonate, they are mainly intended to maintain or raise the pH of the soil and improve its structure. The calcium and magnesium contained in these amendments can be used to correct soil deficiencies in these elements. The farmer, informed by the analysis of his soil's needs, expressed in CaO, can use a whole range of products to carry out liming.

There are five classes of calcium and magnesium soil improvers:

Class I

-**Limestone amendments** (chalk, faluns, maerl, pitch, trez, marl) containing as an essential component calcium carbonate $CaCO_3$ These are raw products.

Class II

-**Magnesian limestone and magnesian soil improvers**, raw products of natural origin, containing magnesium carbonate, whether or not associated with calcium carbonate. Dolomite is a double carbonate of $Ca2^+$ and $Mg2^+$.

Class III

-**Lime,** These are the "baked amendments", comprising :

-agricultural quicklime where the calcium is in the oxide CaO state,

-agricultural slaked lime, obtained after hydration of quicklime,

-Magnesian quicklime from the calcination of calcaro-magnesian rocks,

-magnesian slaked lime resulting from the hydration of magnesian quicklime,

-waste or ash of calcium or magnesian lime, residues from the normal manufacture of calcium or magnesian lime.

Class IV

-**Mixed amendments**, mixtures of "cooked" and "raw" amendments.

Class V

-**Other amendments**, the most important of which is made up of defecation skimmings from sweets, residues from the filtration of sweet juices after carbonation of the latter by a milk of lime.

Conditions of use of limestone improvers :

• in the case of a plantation, a slow action is sought with an amendment having a medium to low carbon solubility. Exceptional case of soils with a pH < 5.8 where it is absolutely necessary to correct the pH quickly by using fast acting amendments.

• on an existing vine and in the case of a situation to be straightened, products with rapid action and high carbon solubility must be used.

• in maintenance on heavy soil, we will carry out contributions every 3 to 5 years with fast-acting soil improvers.

• for maintenance on light soil, applications will be made every 2-3 years using medium-active amendments.

II. HUMIC AMENDMENTS

The presence of organic matter in agricultural soils is essential to ensure good fertility.

Organic matter and humus :

Broadly speaking, three fractions of organic matter can be distinguished: fresh organic matter, matter in the process of transformation and stable organic matter. Fresh organic matter (MOF) is crop residues, recently applied organic fertilizers and soil fauna. ROM represents only 10% of the total organic matter stock. Microbial biomass (bacteria, fungi, microscopic algae) is also fresh organic matter.

Stable organic matter (SOM) represents the largest fraction, with more than 80%. It is this fraction that is commonly referred to as "humus". Humus is therefore the result of the transformation of organic matter into a stable form that will degrade only very slowly. The measurement of the organic matter (OM) content of a soil is based on the determination of organic carbon (C). It is assumed that the OM content equals the organic C content multiplied by a coefficient of 1.72.

Roles of organic matter

Effects on physical properties are best observed in silty and light soils. Organic matter plays a major role in the stability of the structure by increasing the cohesion of the particles. Silt soils with high organic matter contents are less sensitive to threshing and erosion. The stability of the organic matter limits the phenomenon of compaction and facilitates the drying out of the soil. Organic matter also plays a role in the water retention capacity of a soil. It can be assimilated to a sponge that retains water and later releases it again in dry conditions. It is therefore on sandy or shallow soils that high levels of stable organic matter (humus) are of particular interest.

The chemical properties of a soil are also influenced by the organic matter content. The latter, together with clay, is involved in the constitution of the clay-humus complex which contains nutrients that will be taken up by the roots. Clay and humus contents determine the cation exchange capacity (CEC) of a soil. The higher the CEC, the more the soil is able to return nutrients such as magnesium, calcium, potassium and sodium to the plants. Organic matter has a higher CEC than clay. In clay soils, CEC will naturally be higher than in sandy soils, where organic matter plays a greater role in the availability of nutrients. In sandy soils, organic matter therefore helps to slow down the loss of leachable elements such as potassium or magnesium.

Effects on the **biological** properties of a soil are also attributed to organic matter. The application of organic matter stimulates the soil's microbial activity and increases the total microbial biomass.

Desirable levels of organic matter :

Depending on the pedo-climatic context, the desirable levels of organic matter will therefore differ. On loamy soil, 2.2% is considered optimal. In sandy soil, in order to compensate for the low cation exchange capacity, an optimal level of 2.5% should be achieved. On clayey soil, 2% MO may be sufficient. The calcium content of the soil will also determine the importance of the MO requirements. Thus, a calcareous soil must receive a more regular supply of OM.

Increase in humus levels :

Like any stock, the stock of organic matter in a soil can only increase if the inputs are increased. One can also act, but to a much lesser extent, on mineralisation.

- **Acting on the formation of humus** :

The amount of humus formed depends on the amount of organic matter brought to the soil, but also on the nature of the organic matter. The yield of the transformation of organic matter into humus is evaluated by means of the isohumic coefficient **K1**. This translates the percentage of OM that is transformed into humus under the effect of bacterial activity. The higher the coefficient, the higher the proportion of organic matter that is transformed into humus. This is the case for green waste composts, which have K1 coefficients of between 0.5 and 0.7. On the other hand, there are slurries and manure whose organic matter contributes very little to the formation of humus. The type of crops grown on the farm plays an important role in the evolution of humus stocks.

Values of the isohumic coefficient K1

Cattle manure	0,3 à 0,5
Poultry droppings	0,05 à 0,1
Poultry manure	0,1 à 0,15
Compost of green waste	0,5 à 0,7
Household waste compost	0,15 à 0,3
Sewage sludge	0,15 à 0,2
Straw	0,15

Estimated intake of stable MO ("humus") from crop residues

Culture	Quantity of stable OM (in kg/ha)
Winter cereal	550
Buried straw cereal	1050
Corn silage	400
Grain maize	1200
Beetroot with buried leaves	600
Potatoes	150

- **To act on the mineralisation of the humus :**

The humus stock is subject to the phenomenon of mineralisation which, each year, destroys on average 1 to 2% of the total quantity of humus. The quantity of humus mineralised each year **K2** depends mainly on the type of soil and climate. Mineralisation is more important in sandy soils than in silty and clayey soils. On calcareous soils, mineralisation is often less than 1%. The pH of the soil also plays a role in the intensity of the mineralisation. The more acidic the soil, the slower the mineralisation. Cultivation techniques can influence the degradation of humus stocks. For example, deep ploughing will result in a dilution of the humus level. Tillage, by favouring soil fragmentation and aeration, increases the mineralisation phenomenon. In fact, no-tillage or simplified tillage techniques make it possible to limit the destruction of humus.

Humic assessment :

By knowing the quantities formed and mineralised each year, it is possible to establish the humic balance. This balance can be drawn up on the scale of the farm as well as on the scale of each plot.

Regular analysis of the OM content of the plots is an excellent indicator of the evolution of fertility.

Humus sources :

Humigenous amendments are mainly manure, slurry, plant waste, harvest residues, green manure, fermentable household waste and sewage plant sludge.

1. Dung :

Poultry manure is very rich in nitrogen, phosphorus and potassium, while other manure from cattle, sheep, goats and horses is usually deficient in phosphate. Obtaining good manure starts in the barn, where the bedding, well moistened with urine, must be as compact as possible so that anaerobic conditions can be created, guaranteeing less carbon, potassium and nitrogen losses. Because manure is deficient in phosphorus, farmers should supplement it by treating the bedding, for example, with calcium phosphates (natural phosphates or maerl).

2. Liquid manure :

The best way to prepare slurry would be to oxygenate it to maintain aerobic activity in the storage pits, supplement it with phosphates and dilute it.

3. Plant residues and green manure :

The organic matter is crushed, diluted and then moistened. They are put in piles and sufficiently compacted to initiate anaerobic fermentation and heating. After a few days, the heap is taken up again, decompacted so that the desired aerobiosis can take place. It is during this process that soil, limestone and phosphates can be added or the wetting can be controlled. The compost is then left to mature for two to three months. When sufficient space is available, the heap can be taken up again to recompose it and mix the organic materials that are already well composted with those that are not composted enough. When the compost is fully mature, it can be incorporated into the soil or, better still, spread it on the surface and let the soil fauna mix with the mineral particles in the soil.

Green manures are among the interesting humic amendments. The aim is to cultivate a plant that quickly produces a large amount of vegetation that will be buried after crushing. Apart from providing organic matter, the main advantages of green manures are that they do not leave the soil without plant cover and keep the microbial activity of the rhizosphere intact. The role of green manures in improving soil structure is fundamental. The only notable disadvantage of green manures, but it is that of all vegetation, is the evapotranspiration that follows and which risks, more or less, drying out the soil.

II.   MINERAL FERTILISER

The reasoning behind fertilisation is based on...:

• The quantity (dose) to bring.

• The form of the fertilizer.

• The number and dates of the contributions.

• The interest of fertilizers.

The elements necessary for the plant come from the air and the soil. If the soil is abundantly supplied with nutrients, the plants grow well and give high yields.

If the soil is poor in only one of the elements, plant growth is limited and yields are reduced.

To obtain good yields, crops must be provided with the elements that the soil lacks.

Nutritional elements necessary for plant growth:

The vast majority of plants require 16 nutrients to grow:

➢   from the air: Carbon(C) in the form of CO_2 (carbon dioxide) ;

➢   water: Hydrogen (H) and oxygen (O) in the form of water (H_2O) ;

➢   soil and mineral and organic fertilizers :

- basic or major elements (macro elements): Nitrogen (N), phosphorus (P), potassium (K).

- secondary elements: Calcium (Ca), magnesium (Mg), sulphur (S).

- trace elements : Iron (Fe), manganese (Mn), zinc (Zn), copper (Cu), boron (B), molybdenum (Mo), and chlorine (Cl).

Secondary elements and trace elements are usually found in sufficient quantities in the soil and should only be added if a deficiency is found.

The major elements

Nitrogen (N) :

Nitrogen is a major element for plant fertilisation, it is taken from the soil in either nitric (NO_3-) or ammoniacal (NO_4+) form. It has several roles in plant development. It is the engine of plant growth and contributes to the vegetative development of all the aerial parts of the plant, leaves, stems and seed formation, hence its contribution to yield improvement.

Phosphorus (P) :

The role of phosphorus is to strengthen plant resistance and contributes to the growth and development of roots, fruiting and sowing.

Potassium (K) :

Potassium is an element that helps to promote flowering and fruit development. It also has an action of strengthening resistance to disease and cold, limiting evapotranspiration, stiffening the stem, and building up the nutrient reserve (bulbs).

Signs of deficiencies in basic elements: N.P.K.

For nitrogen (N)

- Languid vegetation ;
- light green or yellowish (chlorosis) foliage ;
- Plants of small size.

For phosphorus (P)

- Dark green, bronzed or reddish-stained foliage;
- Slender or poorly formed branches ;
- Not very abundant flowering ;
- Abortion of flowers ;
- Late ripening of the fruit.

For potassium (K)

- Brown necroses at the tip, on the edges and between the veins of the leaves;
- The distance between the knots on the stem becomes smaller,
- The edge of the leaves sometimes curls upwards ;
- Plants susceptible to diseases ;
- Fruits with little sugar and no flavour;
- Poor preservation of root vegetables.

Principle of fertilization

The main objective of fertilisation is to maintain the fertility of the soil in order to satisfy the needs of the crops. The current principles of fertilisation stem from three fundamental laws: the law of restitution to the soil, the law of less than proportional increases and the law of interaction.

The law of restitutions on the ground :

It is based on the compensation of the export of mineral elements by plants by means of refunds to avoid soil depletion. This rule is insufficient for three reasons:

• Some soils are naturally poor in one or more nutrients, so they must be enriched to meet the definition of cultivated soil.

• Generally, soils are exposed to nutrient losses through leaching.

• During certain periods of their vegetative cycle, plants have intense nutrient needs, known as "instant needs", at which time the soil's mobilisable reserves may be insufficient.

The law of less than proportional returns :

When increasing doses of a fertilising element are applied to the soil, yields do not increase proportionally. In fact, the yield increases that are obtained are smaller and smaller as the quantities applied increase.

Thus there is an optimal dose of elements to be brought in, as the maximum dose is not the most economical. Furthermore, fertilisation must take into account the rate of absorption of the elements, the exchange capacity of the soil and the dynamics of the nutrients.

The law of the minimum :

The insufficiency of an assimilable element in the soil reduces the efficiency of the other elements and consequently reduces the crop yield (Liebig's law).

All nutrients must be present in some variable balance with the crop. The major elements (N P K, etc.), must necessarily be present in greater quantities but any Oglio element can act as a limiting factor if there is a deficiency in this element.

The different types of mineral fertilisers :

A distinction is made between **simple** fertilisers, which contain only one nutrient, and **compound** fertilisers, which may contain two or three nutrients. The name of mineral fertilisers is standardised by reference to their three main components: N-P-K.

Simple fertilisers can be nitrogen, phosphate or potassium. Compound fertilisers can be binary (when they contain two elements N-P or P-K or N-K). These letters are usually followed by numbers, representing the respective proportion of these elements.

Industrially produced chemical fertilisers contain a guaranteed minimum amount of nutrients. This quantity is written on the bag.

Simple fertilizers :

■ **ammonium nitrate (UAN, 32% N),** a *liquid polyvalent fertiliser, to be diluted in* water between 5 and 10% depending on the vegetative stage. It is intended for all crops: Cereal crops - potatoes - industrial tomatoes - arboriculture - viticulture.

■ **ammonium sulphate (SA, 21% N), a** *nitrogenous cover fertilizer for* all crops. Cereal crops - market gardening - arboriculture - industrial crops. Also contains a secondary element, sulphur (24%).

■ **Urea (46% N), a** *nitrogenous cover fertilizer, intended for all crops:* Cereal crops - market gardening - arboriculture - viticulture - pulses.

■ **Calcium ammonium nitrate (CAN, 27% N), a** *nitrogenous cover fertilizer, intended for* all crops: Cereal crops - market gardening - arboriculture - viticulture. Also contains two secondary elements: Calcium (7.5%) and magnesium (3.5%).

■Le **sulfazote (26% N),** sulphurous nitrogenous cover fertilizer, intended for all crops: Cereal crops - market gardening - arboriculture - viticulture. Also contains a secondary element: Sulphur (14%).

■ **Simple superphosphate (SSP, 20% P), a** *bottom and cover phosphate fertiliser,* intended for all crops: Cereal crops - pulses - market gardening - arboriculture - industrial crops - fodder crops. Also contains two secondary elements: Calcium (28%) and sulphur (22%), and trace elements : Boron (61ppm), iron (2134 ppm), manganese (27 ppm), zinc (127 ppm), copper (02 ppm).

■ **Triple superphosphate (TSP, 46% P),** *a bottom phosphate fertiliser used before* sowing for cereal crops and pulses. Also contains trace elements Boron (61ppm), iron (3638 ppm), manganese (114 ppm), zinc (170 ppm), copper (05 ppm).

Compound fertilizers :

■ **N.P.K.s (04.20.25)** *is a complex* ternary **nitrogen phosphate potash** *fertiliser.* It contains 4% N, 20% P and 25% K. A bottom fertilizer, it is intended for all perennial crops including viticulture - arboriculture. It also contains a secondary element: sulphur (12%) and trace elements: Boron (29 ppm), iron (2036 ppm), manganese (34 ppm), zinc (173 ppm), copper (02 ppm).

■ **N.P.K.s (10.10.10)** *is a ternary fertilizer* containing 10% N, 10% P and 10% K. It is versatile and is used for market gardening - viticulture - arboriculture as a background fertilizer at the time of sowing and for the different plantations. It is suitable for all types of soil. It also contains micro-nutrients: Boron (30 ppm), iron (1723 ppm).

■ **Chlorinated potassium phosphate nitrogen N.P.K.c (15.15.15)** *is a ternary fertiliser which* contains 15% N, 15% P and 15% K. It is versatile and is used for all market gardening and industrial crops (with the exception of chlorine-sensitive crops) as a background fertiliser at the time of sowing, on non-saline soils with a bleeding capacity.

■ **N.P.K.s (15.15.15)** *is a ternary fertilizer* containing 15% N, 15% P and 15% K. Versatile, it is used for market gardening - viticulture - arboriculture as a background fertilizer at the time of sowing and for the various plantations. It is suitable for all types of soil. It also contains sulphur (8%) and trace elements: Boron (45 ppm), iron (1723 ppm), manganese (30 ppm), zinc (156 ppm), copper (02 ppm).

The notion of fertilizer dose :

The fertilizer dose represents the amount of fertilizer that must be incorporated into the soil to meet the maintenance and production needs of the plants grown there. It should therefore be sufficient to guarantee the harmonious growth of the plant and ensure the expected yield in terms of quantity and quality.

Calculating the fertiliser dose :

Fertilizer unit concentrations in fertilizers are expressed in % (i.e. per 100 kg of commercial product) and crop requirements are given per hectare. The formula for calculating the dose is therefore :

The dose = Plant requirements x 100 kg / The fertiliser dose.

FERTILIZATION EQUIPMENT

Introduction :

Fertilisation aims to maintain and increase the soil's potential by providing fertilisers and amendments. Amendments are used to maintain or correct the physical structure of the soil and are generally mineral products (limestone, marl, agricultural lime...etc.) While fertilisers bring to the soil, and therefore to the plants, the chemical elements essential to their development (nitrogen, phosphoric acid and potassium).

Fertilizers can be of natural or industrial origin. The so-called natural ones are: green manure, manure and slurry. As for industrial fertilisers, they are generally in solid or sometimes powdered (liquid) form.

In any case, the materials must have a high resistance to corrosion, as the products applied are generally very aggressive. Plastics and stainless steel are often used.

I. Centrifugal fertilizer spreaders :

These highly responsive dispensers are mounted or semi-mounted. Mounted units have a capacity of 40 to 2000 litres and semi-mounted units up to 10,000 litres. Spreading widths vary from 9 to 24 metres, while some machines can reach up to 30 metres. The centrifugal distributors are either swing tube or rotary disc type.

I. 1. centrifugal distributor with oscillating tube :

The fertiliser is spread by means of a horizontal conical tube, the oscillation of which throws the fertiliser in a zigzag pattern to the rear in the direction of travel. The fertiliser in the conical hopper falls towards the oscillating tube by gravity. The movement of the tube is driven by the tractor's PTO shaft.

At the bottom of the hopper, an agitator driven by the distribution mechanism ensures the regular flow of fertiliser through adjustable openings that allow the flow rate to be adapted. This flow adjustment consists of two superimposed discs, one fixed and one mobile, which have openings. The rotation of the movable disc, operated by a control lever, changes the flow cross-sections of the fertiliser and thus the flow rate.

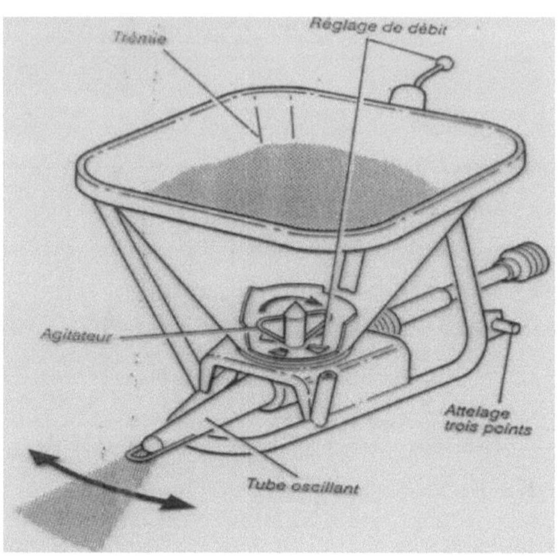

I. 2. Centrifugal distributor with rotating discs :

This type of equipment could be gravity or mechanical distribution. In the case of gravity feed, the equipment includes a hopper, a flow regulation system consisting of adjustable trapdoors located upstream of the discs, and one or two discs with vertical axes. These discs are fitted with more or less radial vanes that eject the fertiliser in an arc behind the machine. In two-disc systems, the rotational speeds are generally opposite.

Mechanical feeding is mainly used on large-capacity feeders, where the hopper cannot be located above the discs. In this case, the fertiliser is conveyed by a longitudinal conveyor to two lateral rubber conveyor belts that feed the discs. The spread rate is adjusted by means of adjustable flaps located upstream of the side belts or by changing the speed of the longitudinal conveyor.

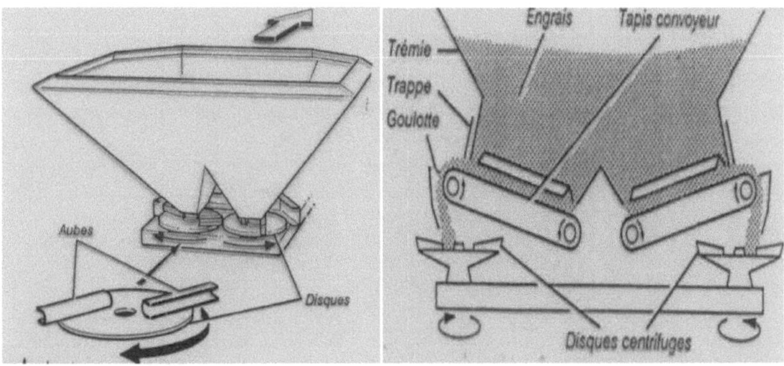

I.3. Pneumatic spool valves :

In this type of distribution, the fertiliser extracted from the hopper is carried in pipes by an air flow produced by a turbine to distribution nozzles or diffusers arranged on a cross-rail. A pneumatic fertiliser spreader consists of a hopper, side feeders, a turbine, pipes and diffusers.

- **The hopper:** With a capacity of 600 to 800 litres, it is in the shape of a truncated pyramid or truncated cone for mounted equipment. It is wider and can reach 5,000 to 6,000 litres for larger appliances.

- Dosing **units:** Dosing units are usually made up of pin or lamella rollers arranged in as many rows as there are pipes to be fed. The flow rate is adjusted manually or automatically (flow rate proportional to the advance) by varying the rotation speed of these rollers using a gearbox or a variator. The drive is carried out by the power take-off for mounted equipment and by the wheels on semi-mounted equipment.

- **Turbine: This** is driven by the PTO and creates an air flow that carries the fertiliser from the spreaders to the diffusers.

- **Piping :** Arranged in parallel on a horizontal ramp, they are made up of rigid or flexible plastic tubes. They are of different lengths, each ending in a diffuser. The length of the ramp varies from 9 to 24 metres.

- **Diffusers:** These are outlet orifices, evenly distributed over the ramp, with a spacing of 35cm to 1m. They are made up either of a ribbed vane forming a deflector, or of a small rotor rotating under the influence of the air flow. Their role is to distribute the fertiliser, with a slight crossover, so as to obtain as uniform a spread as possible.

Pneumatic fertiliser spreaders can be used for localised fertilisation by adapting locating sleeves to the spreaders, which lead the fertiliser to the ground at the desired location.

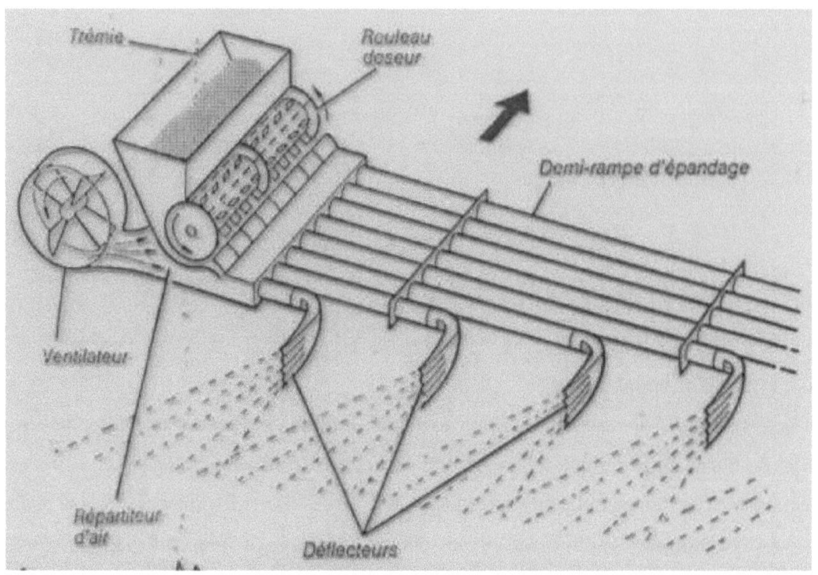

II. Manure spreaders :

Manure has a variable composition and density: its structure ranges from straw manure to compact manure that is difficult to shred. It is commonly spread at 30 to 70 tonnes/ha, which requires the handling of large volumes.

II. 1 Rear spreading manure spreader :

It is a semi-mounted trailer with a moving floor and a rotating shredding and spreading unit, driven by the tractor's power take-off. The capacities vary from 5 to 15m² and the spreading devices can be dismantled in order to use the trailer for other purposes.

• Movable **floor**: this is a movable apron, generally made up of two or four longitudinal chains, connected to each other by profiled metal bars that move at the bottom of the trailer from front to back, dragging the manure. The moving floor (very slowly) and the spreading system are operated by the tractor's power take-off.

• **Spreading organs**: these organs shred and spread the manure pushed by the moving bottom. They consist of one or more horizontal (up to three) or vertical (up to four) rotors of very varied shapes: hedgehogs (rotors fitted with teeth, knives, blades or spade), helical screws, or serrated discs mounted obliquely on a shaft, etc.

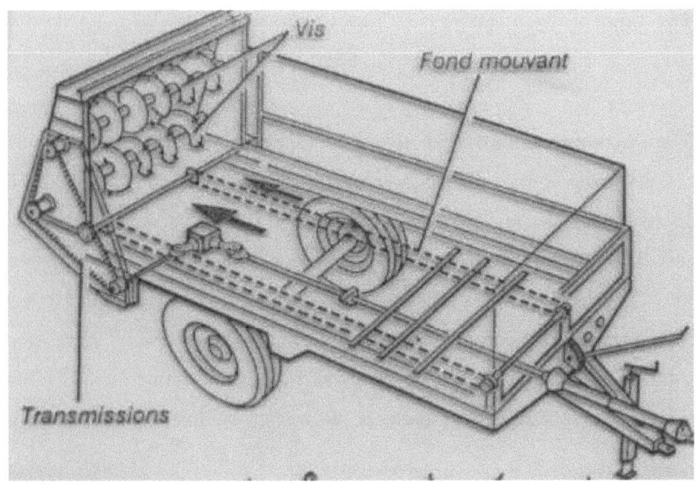

II. 2. side flail spreader :

It is a trailer consisting of a watertight box supporting a longitudinal rotor, equipped with chains and flails. The rotor, driven by the power take-off, can move vertically above the box, to allow a progressive attack of the flails that throw the manure to the side, thanks to an upper deflector.

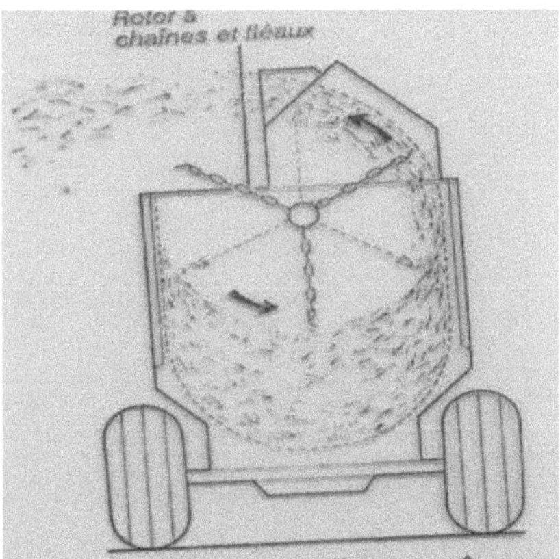

III. Slurry spreader :

Also known as a "slurry tanker", this semi-mounted equipment consists of a tank mounted on a single or double-axle chassis, a pneumatic compressor that provides the energy needed for loading and spreading, and a spreading device.

• **The tank**: it has a capacity of 2 to 10m3 and is made of galvanised steel to resist corrosion. The rear part is fitted with solid bolts and hinges to allow cleaning and removal of deposits.

• **The pneumatic compressor**: this is a rotary compressor driven by the tractor's power take-off. A reversing valve system allows, at the moment of loading, to connect the suction of the compressor with the inside of the tonne. The resulting vacuum allows the slurry to be sucked into the pit using a large diameter filling pipe. For spreading, the operator reverses the flow of the compressor, which pushes the air back into the tonne, in order to expel the slurry through a spreading valve controlled remotely from the tractor driver's seat. An anti-overflow valve system prevents slurry from being sucked in by the compressor. In addition, an air pressure and vacuum relief valve protects the system against overpressure and the risk of bursting.

• Spreading **organs**: the spreading of manure is most often done by sprinkling or sometimes by burying. In the case of sprinkling, the slurry tanker is equipped with a diffuser placed at the outlet of the discharge valve. This diffuser is designed to distribute the slurry as a layer on the ground.

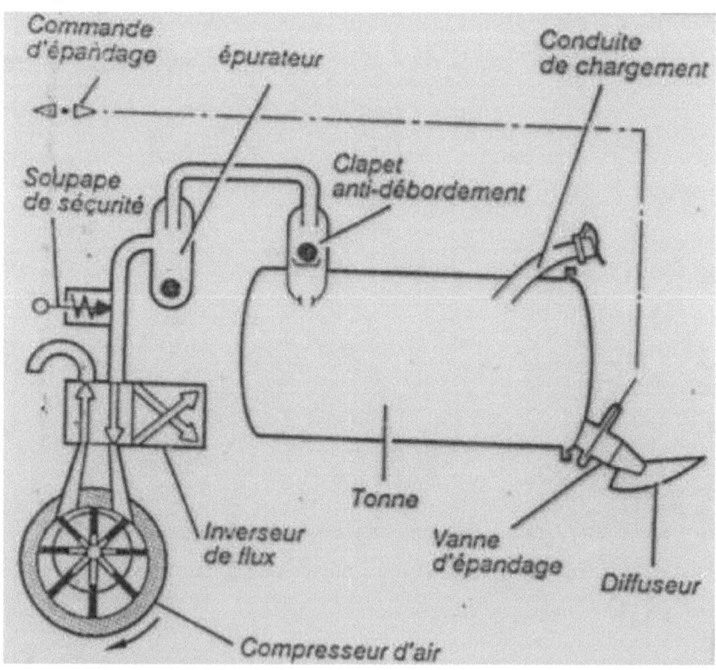

BIBLIOGRAPHICAL REFERENCES

Barthelemy P., Boisgontier D. Lajoux P. , 1992. Choosing soil working tools. ITC F, 194 p. ISBN 2-86492-140-5.

Bassez J., Delouvée R., Habib Z., Corpen, 1997, Bien choisir et mieux utiliser son matériel d'épandage de lisiers ou de fumiers. Brochure produced by the Spreading sub-group, TRAME-BCMA, FNCUMA, Agence Seine Normandie, Étude CORPEN, 55 p.

Candelon P., 1981. Agricultural machinery Volume 1: Soil preparation and fertilisation equipment. Ed. J-b Baillière, 218 p.

Cédra C. , 1997. Fertilisation and crop treatment equipment. Ed. CEMAGREF, FNCUMA, ITCF, Lavoisier Tec and Doc, 343 p. ISBN 2-85362-458-7.

Cédra C. , 1993. Tillage, sowing and planting materials. CEMAGREF, FNCUMA, ITCF, Lavoisier Tec and Doc, 384 pp. ISBN 2-85362-348-3

Cédra C. , 1991. Illustrated glossary of agricultural machinery and equipment. Ed. Cemagref DICOVA and Lavoisier tec et Doc, 350p.

CNEEMA, 1963. Tractor and agricultural machinery - Master's book - Volume 2. Ed. Antony (Seine): Centre national d'études et d'expérimentation de machinisme agricole, 264 p.

Hénin S., Féodoroff A., Gras R., 1969. The cultural profile: the physical state of the soil and its agronomic consequences. Ed. Paris Masson, 332p.

Hénin S., Féodoroff A., Gras R., Monnier G., 1960. The cultural profile. Principles of soil physics. Editions du Seuil, SELA Paris, 320p.

Lerat P. 2015. Agricultural machinery operation and maintenance. Ed. Agriculture d'Aujourd'hui, 440 p.

Oestges O. 1994. The mechanisation of agricultural work: Volume 1. Ed. Les presses agronomiques de Gembloux, 248 p. ISBN 2-87016-044-5.

Soltner D., 2005. The basics of plant production. Volume 1: The soil and its improvement. Ed. Sciences et Techniques Agricoles, 472p. ISBN 2-907710-00-1.

Buy your books fast and straightforward online - at one of world's fastest growing online book stores! Environmentally sound due to Print-on-Demand technologies.

Buy your books online at
www.morebooks.shop

Kaufen Sie Ihre Bücher schnell und unkompliziert online – auf einer der am schnellsten wachsenden Buchhandelsplattformen weltweit! Dank Print-On-Demand umwelt- und ressourcenschonend produzi ert.

Bücher schneller online kaufen
www.morebooks.shop

KS OmniScriptum Publishing
Brivibas gatve 197
LV-1039 Riga, Latvia
Telefax: +371 686 204 55

info@omniscriptum.com
www.omniscriptum.com